M000268266

GACE 013 Middle Grades Mathematics

Teacher Certification Exam

By: Sharon Wynne, M.S.
Southern Connecticut State University

"And, while there's no reason yet to panic, I think it's only prudent that we make preparations to panic."

XAMonline, INC.

Boston

Library of Congress Cataloging-in-Publication Data

Wynne, Sharon A.
    GACE: Middle Grades Mathematics 013 Teacher Certification / Sharon A. Wynne. -2nd ed.
    ISBN: 978-1-58197-345-7
    1. Middle Grades Mathematics 013   2. Study Guides.   3. GACE
    4. Teachers' Certification & Licensure.   5. Careers

Disclaimer:

The opinions expressed in this publication are the sole works of XAMonline and were created independently from the National Education Association, Educational Testing Service, or any State Department of Education, National Evaluation Systems or other testing affiliates.

Between the time of publication and printing, state specific standards as well as testing formats and website information may change that is not included in part or in whole within this product. Sample test questions are developed by XAMonline and reflect similar content as on real tests; however, they are not former tests. XAMonline assembles content that aligns with state standards but makes no claims nor guarantees teacher candidates a passing score. Numerical scores are determined by testing companies such as NES or ETS and then are compared with individual state standards. A passing score varies from state to state.

Printed in the United States of America œ-1

GACE: Middle Grades Mathematics 013
ISBN: 978-1-58197-345-7

Table of Contents

SUBAREA III.   PATTERNS, ALGEBRA, AND FUNCTIONS

COMPETENCY 7.0   UNDERSTAND PATTERNS, RELATIONS, AND FUNCTIONS

COMPETENCY 8.0   UNDERSTAND THE PROPERTIES AND TECHNIQUES OF ALGEBRAIC OPERATIONS

Great Study and Testing Tips!

What to study in order to prepare for the subject assessments is the focus of this study guide but equally important is *how* you study.

You can increase your chances of truly mastering the information by taking some simple, but effective steps.

Study Tips:

1. <u>Some foods aid the learning process</u>. Foods such as milk, nuts, seeds, rice, and oats help your study efforts by releasing natural memory enhancers called CCKs (*cholecystokinin*) composed of *tryptopha*n, *choline*, and *phenylalanine*. All of these chemicals enhance the neurotransmitters associated with memory. Before studying, try a light, protein-rich meal of eggs, turkey, and fish. All of these foods release the memory enhancing chemicals. The better the connections, the more you comprehend.

Likewise, before you take a test, stick to a light snack of energy boosting and relaxing foods. A glass of milk, a piece of fruit, or some peanuts all release various memory-boosting chemicals and help you to relax and focus on the subject at hand.

2. <u>Learn to take great notes</u>. A by-product of our modern culture is that we have grown accustomed to getting our information in short doses (i.e. TV news sound bites or USA Today style newspaper articles.)

Consequently, we've subconsciously trained ourselves to assimilate information better in <u>neat little packages</u>. If your notes are scrawled all over the paper, it fragments the flow of the information. Strive for clarity. Newspapers use a standard format to achieve clarity. Your notes can be much clearer through use of proper formatting. A very effective format is called the *"Cornell Method."*

> Take a sheet of loose-leaf lined notebook paper and draw a line all the way down the paper about 1-2" from the left-hand edge.

> Draw another line across the width of the paper about 1-2" up from the bottom. Repeat this process on the reverse side of the page.

Look at the highly effective result. You have ample room for notes, a left hand margin for special emphasis items or inserting supplementary data from the textbook, a large area at the bottom for a brief summary, and a little rectangular space for just about anything you want.

3. <u>**Get the concept then the details.**</u> Too often we focus on the details and don't gather an understanding of the concept. However, if you simply memorize only dates, places, or names, you may well miss the whole point of the subject.

A key way to understand things is to put them in your own words. If you are working from a textbook, automatically summarize each paragraph in your mind. If you are outlining text, don't simply copy the author's words.

Rephrase them in your own words. You remember your own thoughts and words much better than someone else's, and subconsciously tend to associate the important details to the core concepts.

4. <u>**Ask Why?**</u> Pull apart written material paragraph by paragraph and don't forget the captions under the illustrations.

Example: If the heading is "Stream Erosion", flip it around to read "Why do streams erode?" Then answer the questions.

If you train your mind to think in a series of questions and answers, not only will you learn more, but it also helps to lessen the test anxiety because you are used to answering questions.

5. <u>**Read for reinforcement and future needs.**</u> Even if you only have 10 minutes, put your notes or a book in your hand. Your mind is similar to a computer; you have to input data in order to have it processed. *By reading, you are creating the neural connections for future retrieval.* The more times you read something, the more you reinforce the learning of ideas.

Even if you don't fully understand something on the first pass, *your mind stores much of the material for later recall.*

6. <u>**Relax to learn so go into exile.**</u> Our bodies respond to an inner clock called biorhythms. Burning the midnight oil works well for some people, but not everyone.

If possible, set aside a particular place to study that is free of distractions. Shut off the television, cell phone, pager and exile your friends and family during your study period.

If you really are bothered by silence, try background music. Light classical music at a low volume has been shown to aid in concentration over other types. Music that evokes pleasant emotions without lyrics are highly suggested. Try just about anything by Mozart. It relaxes you.

7. <u>Use arrows not highlighters</u>. At best, it's difficult to read a page full of yellow, pink, blue, and green streaks. Try staring at a neon sign for a while and you'll soon see that the horde of colors obscure the message.

A quick note, a brief dash of color, an underline, and an arrow pointing to a particular passage is much clearer than a horde of highlighted words.

8. <u>Budget your study time</u>. Although you shouldn't ignore any of the material, *allocate your available study time in the same ratio that topics may appear on the test.*

Testing Tips:

1. Get smart, play dumb. **Don't read anything into the question.** Don't make an assumption that the test writer is looking for something else than what is asked. Stick to the question as written and don't read extra things into it.

2. Read the question and all the choices *twice* before answering the question. You may miss something by not carefully reading, and then re-reading both the question and the answers.

If you really don't have a clue as to the right answer, leave it blank on the first time through. Go on to the other questions, as they may provide a clue as to how to answer the skipped questions.

If later on, you still can't answer the skipped ones . . . *Guess.* The only penalty for guessing is that you *might* get it wrong. Only one thing is certain; if you don't put anything down, you will get it wrong!

3. Turn the question into a statement. Look at the way the questions are worded. The syntax of the question usually provides a clue. Does it seem more familiar as a statement rather than as a question? Does it sound strange?

By turning a question into a statement, you may be able to spot if an answer sounds right, and it may also trigger memories of material you have read.

4. Look for hidden clues. It's actually very difficult to compose multiple-foil (choice) questions without giving away part of the answer in the options presented.

In most multiple-choice questions you can often readily eliminate one or two of the potential answers. This leaves you with only two real possibilities and automatically your odds go to Fifty-Fifty for very little work.

5. Trust your instincts. For every fact that you have read, you subconsciously retain something of that knowledge. On questions that you aren't really certain about, go with your basic instincts. **Your first impression on how to answer a question is usually correct.**

6. Mark your answers directly on the test booklet. Don't bother trying to fill in the optical scan sheet on the first pass through the test.

7. Watch the clock! You have a set amount of time to answer the questions. Don't get bogged down trying to answer a single question at the expense of 10 questions you can more readily answer.

SUBAREA I.   **NUMBERS AND OPERATIONS**

COMPETENCY 1.0   UNDERSTAND THE STRUCTURE OF THE BASE TEN NUMERATION SYSTEM AND NUMBER THEORY

Skill 1.1   **Analyze the structure of the base ten number system (e.g., decimal and whole number place value)**

The number system we use is based on ten digits (0−9). Each digit has a place value in a number. Each place value tells the value of a digit in a number.

Whole numbers are the counting numbers along with zero. For example, 0, 1, 2, 3, 4, …

Decimals are portions of ten (deci = part of ten). To find the decimal equivalent of a fraction, use the denominator to divide the numerator as shown in the following example.

<u>Example:</u> Find the decimal equivalent of $\dfrac{7}{10}$.

Since 10 cannot divide into 7 evenly

$$\frac{7}{10} = 0.7$$

Whole Number Place Values are where the digits fall to the left of the decimal point.

Consider the number 792. We can assign a place value to each digit.

Reading from left to right, the first digit (7) represents the hundreds place. The hundreds place tells us how many sets of one hundred the number contains. Thus, there are 7 sets of one hundred in the number 792.

The second digit (9) represents the tens place. The tens place tells us how many sets of ten the number contains. Thus, there are 9 sets of ten in the number 792.

The last digit (2) represents the ones place. The ones place tells us how many ones the number contains. Thus, there are 2 ones in the number 792.

Therefore, there are 7 sets of 100, plus 9 sets of 10, plus 2 ones in the number 792.

Decimal Place Value is where the digits fall to the right of the decimal point.

More complex numbers have additional place values to both the left and right of the decimal point. Consider the number 4.873.

Reading from left to right, the first digit, 4, is in the ones place and tells us the number contains 4 ones.

After the decimal, the 8 is in the tenths place and tells us the number contains 8 tenths.

The 7 is in the hundredths place and tells us the number contains 7 hundredths.

The fourth digit, 3, is in the thousandths place and tells us the number contains 3 thousandths.

Each digit to the left of the decimal point increases progressively in powers of ten. Each digit to the right of the decimal point decreases progressively in powers of ten.

Example: 12,345.6789 occupies the following powers of ten positions:

10^4	10^3	10^2	10^1	10^0	0	10^{-1}	10^{-2}	10^{-3}	10^{-4}
1	2	3	4	5	.	6	7	8	9

Example: What is the value of the 4 and the 9 in the number 24,016,328.0795?

The 4 is in the millions place, so its value is 4,000,000. The 9 is in the thousandths place, so its value is 0.009.

Skill 1.2　**Demonstrate knowledge of the characteristics of whole numbers (e.g., prime/composite, divisibility)**

Whole numbers have characteristics that make them unique. For instance, they can be divisible by two numbers (factors) or by several factors. Every whole number except 1 is either a prime number or a composite number.
Factors are whole numbers that can be multiplied together to get another whole number.

Prime numbers are whole numbers greater than 1 that have only two factors – 1 and the number itself. Examples of prime numbers are 2, 3, 5, 7, 11, 13, 17, and 19. Note that 2 is the only even prime number.

Composite numbers are whole numbers that have factors other than 1 and the number itself. For example, 9 is composite because 3 is a factor in addition to 1 and 9. The number 70 is also composite because, besides the factors of 1 and 70, the numbers 2, 5, 7, 10, 14, and 35 are also all factors.

Remember that the number 1 is neither prime nor composite.

The following are some rules for **divisibility:**

a. A number is divisible by 2 if that number is even (which means it ends in 0, 2, 4, 6, or 8).

Example: 1,354 ends in 4, so it is divisible by 2. 240,685 ends in a 5, so it is not divisible by 2.

b. A number is divisible by 3 if the sum of its digits is evenly divisible by 3.

Example: The sum of the digits of 964 is 9 + 6 + 4 = 19. Since 19 is not divisible by 3, neither is 964. The sum of the digits of 86,514 is 8 + 6 + 5 + 1 + 4 = 24. Since 24 is divisible by 3, 86,514 is also divisible by 3.

c. A number is divisible by 4 if the number in its last 2 digits is divisible by 4.

Example: The number 113,336 ends with the number 36 in the last 2 places. Since 36 is divisible by 4, then 113,336 is also divisible by 4.

The number 135,627 ends with the number 27 in the last 2 places. Since 27 is not divisible by 4, then 135,627 is also not divisible by 4.

d. A number is divisible by 5 if the number ends in either a 5 or a 0.

Example: 225 ends with a 5 so it is divisible by 5. 2,358 is not divisible by 5 because its last digit is an 8, not a 5 or a 0.

e. A number is divisible by 6 if the number is even and the sum of its digits is evenly divisible by 3.

Example: 4,950 is an even number and its digits add to 18. (4 + 9 + 5 + 0 = 18) Since the number is even and the sum of its digits is 18 (which is divisible by 3), then 4,950 is divisible by 6. 326 is an even number, but its digits add up to 11. Since 11 is not divisible by 3, then 326 is not divisible by 6.

f. A number is divisible by 8 if the number in its last 3 digits is evenly divisible by 8.

Example: The number 113,336 ends with the 3-digit number 336 in the last three places. Since 336 is divisible by 8, then 113,336 is also divisible by 8.

The number 465,627 ends with the number 627 in the last three places. Since 627 is not evenly divisible by 8, then 465,627 is also not divisible by 8.

g. A number is divisible by 9 if the sum of its digits is evenly divisible by 9.

Example: The sum of the digits of 874 is 8 + 7 + 4 = 19. Since 19 is not divisible by 9, neither is 874. The digits of 116,514 are 1 + 1 + 6 + 5 + 1 + 4 = 18. Since 18 is divisible by 9, 116,514 is also divisible by 9.

h. A number is divisible by 10 if the number ends in the digit 0.

Example: 305 ends with a 5 so it is not divisible by 10. The number 2,030,270 is divisible by 10 because its last digit is a 0.

Example: Is the number 387 prime or composite? If the number is composite, list the factors.

The number 387 is composite. Its factors are 1, 3, 9, 43, 129, 387.

Skill 1.3 **Apply the Fundamental Theorem of Arithmetic to determine the prime factorization of numbers**

The Fundamental Theorem of Arithmetic states that every composite (non-prime) number can be written as a product of primes in one, and only one way.

The **Prime factorization** of a number is when a number is written as the product of prime numbers or **prime factors**. To get the prime factors of a number, the number is factored into any 2 factors. The resulting factors are checked to see if either can be factored again. Continue factoring until all remaining factors are prime. This is the list of prime factors. Regardless of what way the original number was factored, the final list of prime factors will always be the same.

Example: Factor 30 into prime factors.
Factor 30 into any 2 factors, e.g. 5 and 6.

$5 \cdot 6$	Now factor the 6.
$5 \cdot 2 \cdot 3$	These are all prime factors.

Or factor 30 into any 2 factors, 3 and 10.

$3 \cdot 10$	Now factor the 10.
$3 \cdot 2 \cdot 5$	These are the same prime factors even though the original factors were different.

Example: Factor 240 into prime factors.
Factor 240 into any 2 factors.

$24 \cdot 10$	Now factor both 24 and 10.
$4 \cdot 6 \cdot 2 \cdot 5$	Now factor both 4 and 6.
$2 \cdot 2 \cdot 2 \cdot 3 \cdot 2 \cdot 5$	These are prime factors.

This can also be written as $2^4 \cdot 3 \cdot 5$.

Skill 1.4   **Use principles of number theory (e.g., greatest common factor, least common multiple) to solve problems in applied contexts**

The **greatest common factor (GCF)** is the greatest number that is a factor of all the numbers given in a series. The GCF can be no greater than the smallest number given in the series. To find the GCF, list all possible factors of the least number given (including the number itself). Starting with the greatest factor (which is the number itself), determine if it is also a factor of all the other given numbers. If so, that is the GCF. If that factor does not work, try the same method on the next factor. Continue until a common factor is found. If no other number is a common factor, then the GCF will be the number 1. Note: There can be other common factors besides the GCF.

Example: Find the GCF of 12, 20, and 36.

The smallest number in the problem is 12. The factors of 12 are 1, 2, 3, 4, 6, and 12. The greatest factor is 12, but it does not divide evenly into 20. Neither does 6, but 4 will divide into both 20 and 36 evenly. Therefore, 4 is the GCF.

Example: Find the GCF of 14 and 15.

Factors of 14 are 1, 2, 7, and 14. The greatest factor is 14, but it does not divide evenly into 15. Neither does 7 or 2. The only factor that is common to both 14 and 15, is 1, therefore 1 is the GCF.

The **least common multiple (LCM)** of a group of numbers is the smallest number into which all of the given numbers will divide evenly. The least common multiple will always be the largest of the given numbers or a multiple of the largest number. You can also find the LCM of numbers by listing all the multiples of the numbers until you find a match.

Example: Find the LCM of 20, 30, and 40.

The greatest number given is 40, but 30 will not divide evenly into 40. The next multiple of 40 is 80 (2 x 40), but 30 will not divide evenly into 80 either. The next multiple of 40 is 120. 120 is divisible by both 20 and 30, so 120 is the LCM.

Example: Find the LCM of 96, 16, and 24.

The greatest number is 96. The number 96 is divisible by both 16 and 24, so 96 is the LCM.

COMPETENCY 2.0   UNDERSTAND DIFFERENT REPRESENTATIONS OF REAL NUMBERS

Skill 2.1   Identify and analyze a variety of models for representing numbers (e.g., fraction strips, diagrams, number lines)

Even though middle school students are typically ready to approach mathematics in abstract ways, some of them still require concrete referents such as manipulatives. It is useful to keep in mind that the developmental stages of individuals vary. In addition, different people have different learning styles, some tending more towards the visual and others relatively verbal. Research has shown that learning is most effective when information is presented through multiple modalities or representations. Most mathematics textbooks now use this multi-modal approach.

Manipulatives are materials that students can physically handle and move. Manipulatives allow students to understand mathematic concepts by allowing them to see concrete examples of abstract processes. Manipulatives are attractive to students because they appeal to the students' visual and tactile senses. Available for all levels of math, manipulatives are useful tools for reinforcing operations and concepts. They are not, however, a substitute for the development of sound computational skills.

Example: Using 12 tiles, build rectangles of equal area but different perimeters.

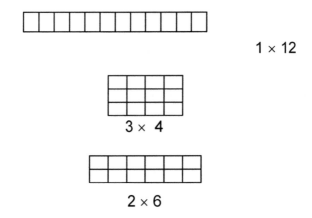

1×12

3×4

2×6

Example: The shaded region below represents 46 out of 100 or 0.46 or $\frac{46}{100}$ or 46%.

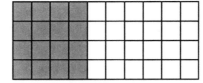

Fraction Strips:

1
2
4
10
5
3
12

Number Lines:

Diagrams:

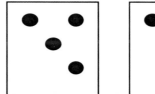

Two groups of four equal eight or 2 x 4 = 8 shown in picture form.

Example: What does the diagram below represent?

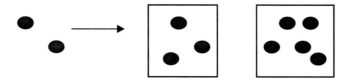

Adding two objects to three or 2 + 3 = 5.

Skill 2.2 **Demonstrate knowledge of equivalency among different representations of numbers (e.g., common and decimal fractions, percents, roots, scientific notation)**

Mathematical manipulation often involves converting real numbers from one form to another for convenience of comparison or ease of interpretation.

To convert a **fraction** to a **decimal**, simply divide the numerator (top) by the denominator (bottom). Use long division if necessary. Note: decimal comes from deci- or part of ten.

Example: Find the decimal equivalent of $\frac{7}{10}$.

$$
\begin{array}{r}
0.7 \\
10\overline{)7.0} \\
\underline{70} \\
00
\end{array}
$$

Since 10 cannot divide into 7 evenly, put a decimal point in the answer row on top; put a zero behind 7 to make it 7.0. Continue the division process. If a remainder occurs, put a zero by the last digit of the remainder and continue the division.

Thus $\frac{7}{10} = 0.7$

Writing zero before the decimal point ensures that the decimal point is emphasized.

Example: Find the decimal equivalent of $\frac{7}{125}$.

$$
\begin{array}{r}
0.056 \\
125\overline{)7.000} \\
\underline{625} \\
750 \\
\underline{750} \\
0
\end{array}
$$

If a decimal has a fixed number of digits, the decimal is said to be terminating. To write such a decimal as a fraction,

multiply by 1 in the form of a fraction (e.g. $\frac{10}{10}, \frac{100}{100}, \frac{1,000}{1,000}$) to get rid of the decimal point.

<u>Example</u>: Convert 0.056 to a fraction.

Multiply 0.056 by $\dfrac{1,000}{1,000}$ to get rid of the decimal point

$$0.056 \cdot \dfrac{1,000}{1,000} = \dfrac{56}{1,000} = \dfrac{7}{125}$$

If a decimal continues forever by repeating a string of digits, the decimal is said to be repeating. To write a repeating decimal as a fraction, follow these steps.

1. Let x = the repeating decimal.
   (e.g. $x = 0.716716716...$)
2. Multiply x by the multiple of ten that will move the decimal just to the right of the repeating block of digits.
   (e.g. $1,000x = 716.716716...$)
3. Subtract the first equation from the second.
   (e.g. $1,000x - x = 716.716.716... - 0.716716...$)
4. Simplify and solve this equation. The repeating block of digits will subtract out.
   (e.g. $999x = 716$ so $x = \dfrac{716}{999}$)

The solution will be the fraction for the repeating decimal.

A **decimal** can be converted to a **percent** by multiplying by 100% or moving the decimal point two places to the right. A **percent** can be converted to a **decimal** by dividing by 100% or moving the decimal point two places to the left.

<u>Example</u>: Convert the following decimals into percents.
   0.375 = 37.5%
   0.7 = 70%
   0.04 = 4%
   3.15 = 315 %

<u>Example</u>: Convert the following percents into decimals.
   84% = 0.84
   3% = 0.03
   60% = 0.6
   110% = 1.1
   $\frac{1}{2}$ % = 0.5% = 0.005

A **percent** can be converted to a **fraction** by dividing by 100% and reducing to simplest terms.

Example: Convert the following percents into fractions.

$32\% = \frac{32}{100} = \frac{8}{25}$

$6\% = \frac{6}{100} = \frac{3}{50}$

$111\% = \frac{111}{100} = 1\frac{11}{100}$

$10\% = \frac{10}{100} = \frac{1}{10}$

The **percentage** of a number can be found by converting the percentage into decimal form, then multiplying the decimal by the number.

Example: Find 23% of 1,000.

$23\% = 0.23$

$0.23 \times 1,000 = 230$

A radical is an integral root of a real number. The nth root of a real number a is written as $\sqrt[n]{a}$ where n is a positive integer greater than 1. Familiar radicals include square roots (e.g. $\sqrt{6}$) and cube roots (e.g. $\sqrt[3]{5}$).

A radical may also be expressed using a rational exponent in the following way:

$$\sqrt[n]{a} = a^{\frac{1}{n}}$$

Example: $\sqrt{5} = 5^{\frac{1}{2}}$; $\sqrt[5]{7} = 7^{\frac{1}{5}}$

To simplify **square roots**, group like factors in pairs. For each of these groups, put one of the factors outside the radical. Any factors that cannot be combined in groups should be multiplied together and left inside the radical.

The index number of a radical is the little number on the front of the radical. For a cube root, the index is 3. If no index appears, then the index is 2 and is the square root.

Example: Simplify $\sqrt[3]{432}$

$$\sqrt[3]{432} = \sqrt[3]{3 \times 2 \times 2 \times 2 \times 3 \times 3 \times 2} = \sqrt[3]{3^3 2^4} = 3 \cdot 2 \cdot \sqrt[3]{2} = 6 \cdot \sqrt[3]{2}$$

When adding and subtracting expressions involving roots, like radicals can be combined.

Example: Simplify $\sqrt{288} + \sqrt{50} + \sqrt{12}$

$$\sqrt{288} + \sqrt{50} + \sqrt{12} = \sqrt{2 \times 144} + \sqrt{25 \times 2} + \sqrt{4 \times 3}$$
$$= 12\sqrt{2} + 5\sqrt{2} + 2\sqrt{3} = 17\sqrt{2} + 2\sqrt{3}$$

Scientific notation is a more convenient method for writing very large and very small numbers. It employs two factors. The first factor is a number between -10 and 10. The second factor is a power of 10. This notation is a shorthand way to express large numbers (like the weight of 100 freight cars in kilograms) or small numbers (like the weight of an atom in grams).

$10^n = (10)^n$ Ten multiplied by itself n times.

$10^6 = 1,000,000$ (mega)

$10^3 = 10 \times 10 \times 10 = 1000$ (kilo)

$10^2 = 10 \times 10 = 100$ (hecto)

$10^1 = 10$ (deca)

$10^0 = 1$ Any nonzero number raised to power of zero is 1.

$10^{-1} = 1/10$ (deci)

$10^{-2} = 1/100$ (centi)

$10^{-3} = 1/1000$ (milli)

$10^{-6} = 1/1,000,000$ (micro)

Scientific notation format. Convert a number to a form of $b \times 10^n$ where $-10 < b < 10$ and n is an integer.

Example: Write 356.73 in various forms.

$$356.73 = 3567.3 \times 10^{-1} \quad (1)$$
$$= 35673 \times 10^{-2} \quad (2)$$
$$= 35.673 \times 10^1 \quad (3)$$
$$= 3.5673 \times 10^2 \quad (4)$$
$$= 0.35673 \times 10^3 \quad (5)$$

Only (4) is written in proper scientific notation format.

Example: Write 46,368,000 in scientific notation.

1. Introduce a decimal point. 46,368,000 = 46,368,000.0

2. Move the decimal place to **left** until only one nonzero digit is in front of it, in this case between the 4 and 6.

3. Count the number of digits the decimal point moved, in this case 7. This is the nth power of ten and is **positive** because the decimal point moved **left**.

Therefore, $46,368,000 = 4.6368 \times 10^7$

Example: Write 0.00397 in scientific notation.

1. The decimal point is already in place.

2. Move the decimal point to the **right** until there is only one nonzero digit in front of it, in this case between the 3 and 9.

3. Count the number of digits the decimal point moved, in this case 3. This is the nth power of ten and is **negative** because the decimal point moved **right**.

Therefore, $0.00397 = 3.97 \times 10^{-3}$.

Example: Evaluate $\dfrac{3.22 \times 10^{-3} \times 736}{0.00736 \times 32.2 \times 10^{-6}}$

Since we have a mixture of large and small numbers, convert each number to scientific notation:

$$736 = 7.36 \times 10^2$$

$$0.00736 = 7.36 \times 10^{-3}$$

$$32.2 \times 10^{-6} = 3.22 \times 10^{-5} \quad \text{thus we have,}$$

$$\frac{3.22 \times 10^{-3} \times 7.36 \times 10^2}{7.36 \times 10^{-3} \times 3.22 \times 10^{-5}}$$

$$= \frac{3.22 \times 7.36 \times 10^{-3} \times 10^2}{7.36 \times 3.22 \times 10^{-3} \times 10^{-5}}$$

$$= \frac{3.22 \times 7.36}{7.36 \times 3.22} \times \frac{10^{-1}}{10^{-8}}$$

$$= \frac{3.22 \times 7.36}{3.22 \times 7.36} \times 10^{-1} \times 10^8$$

$$= \frac{23.6992}{23.6992} \times 10^7$$

$$= 1 \times 10^7 = 10,000,000$$

Skill 2.3   Distinguish between rational and irrational numbers

Rational numbers can be expressed as the ratio of two integers, $\frac{a}{b}$ where $b \neq 0$, for example $\frac{2}{3}$, $-\frac{4}{5}$, $5 = \frac{5}{1}$.

Rational numbers include integers, fractions and mixed numbers, and terminating and repeating decimals. Every rational number can be expressed as a repeating or terminating decimal and can be shown on a number line.

The fraction $\frac{4}{5}$ is a rational number because it is of the form $\frac{a}{b}$ where "a" and "b" are integers and "b" is not zero.

The rational number $\frac{4}{5}$ can be written as 0.8, a terminating decimal by dividing the denominator into the numerator.

The rational number $\frac{7}{11}$ can be written as 0.636363... a non-terminating repeating decimal.

Irrational numbers are real numbers that cannot be written as the ratio of two integers. These are infinite non-repeating and non-terminating decimals.
<u>Example</u>: $\sqrt{5} = 2.236...$ pi $= \prod = 3.1415927...$

The set of rational numbers and the set of irrational numbers have no elements in common and are said to be mutually exclusive.

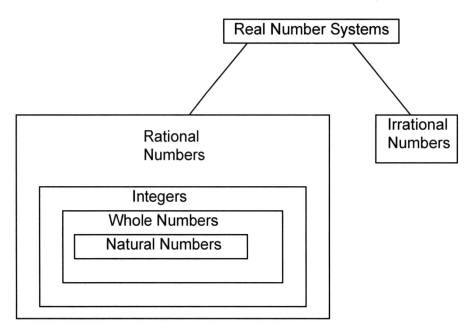

The rational numbers and irrational numbers are subsets of the real number system.

Example: Is $\sqrt{2}$ rational or irrational? Explain your answer.

The number $\sqrt{2}$ is irrational. The square root of 2 is 1.4142..., a non-terminating, non-repeating decimal, which is an irrational number.

Skill 2.4   Compare, order, and round different representations of numbers

When comparing and ordering numbers in different representations, it is convenient to manipulate the numbers so that they are all the same representation. For example, if a student is asked to compare three numbers: a fraction, a decimal, and a square root, encourage the child to convert the numbers to decimals and then compare.

Example: Complete the number sentence.

$$\frac{7}{8} \underline{\hspace{1cm}} 0.88$$

To solve, convert $\frac{7}{8}$ to a decimal. $\frac{7}{8} = 0.875$

Now compare the two decimals: $0.875 < 0.88$.

So, $\frac{7}{8} < 0.88$

Example: Order the following numbers: 8, $\sqrt{60}$, $\frac{125}{16}$, 8.008.

$$\sqrt{60} \approx 7.746, \quad \frac{125}{16} = 7.8125$$

$$\sqrt{60}, \quad \frac{125}{16}, \quad 8, \quad 8.008$$

When rounding to a given place value, it is necessary to look at the number in the next smaller place value. If that number is 5 or greater, the number being rounded is increased by one and all numbers to the right are changed to zero. If the number is less than 5, the number being rounded stays the same and all numbers to the right are changed to zero.

One method of rounding measurements can require an additional step. First, the measurement must be converted to a decimal number. Then the rules for rounding applied.

<u>Example:</u> Round the measurements to the given units.

1 foot 7 inches → feet
5 pound 6 ounces → pounds
$5\frac{9}{16}$ inches → inches

Convert each measurement to a decimal number. Then apply the rules for rounding.

1 foot 7 inches = $1\frac{7}{12}$ ft = 1.58333 ft, round up to 2 ft

5 pounds 6 ounces = $5\frac{6}{16}$ pounds = 5.375 pound, round to 5 pounds

$5\frac{9}{16}$ inches = 5.5625 inches, round up to 6 inches

Skill 2.5 **Select and use different representations of numbers to solve problems in applied contexts**

Different representations of numbers are used for different situations. For example, fractions are often used for certain measurements, decimals are used for currency, and percents are often used for probability.

Fractions are used to represent smaller portions of a larger quantity. They are often used in measuring physical quantities and in dividing whole quantities into equal parts.

Examples of situations where fractions are appropriate are:

Cooks measuring for recipes: 1/2 cup of flour

Carpenters measuring boards: 2 ¾" x 4 ¼" x 36 ¼"

Architects creating models and drafts of rooms and structures: drawn to ¼ scale

Builders apportioning land into lots: ¾ acre for each lot

Servers preparing foods such as pizza or pie: each person will get 1/8 of the pie

Decimals are based on the number 10 and are used to represent quantities between 0 and 1. Sometimes, they are more appropriate than fractions.

Examples of situations where decimals are appropriate:

> Using the metric system
> Expressing U.S. money
> Taking body temperatures
> Measuring distances, time, and gasoline mileage

Percents are used to denote how large one quantity is in relation to another.

Examples of situations where percents are appropriate:

> Commissions
> Discounts
> Taxes
> Tips
> Interest on credit cards, loans, investments
> Sales mark-up
> Nutrition (RDA)
> Sports statistics
> Portions of populations for statistical purposes

Roots and **exponents** are frequently used in finance and science.

Examples of situations where roots and exponents are used:

> Amortization of loans
> Describing acid rain, measuring pH
> Earthquakes (Richter scale)
> The decibel level of sound
> Brightness of stars

Scientific notation uses exponents to express very large numbers easily.

Example of scientific notation:

$3{,}725{,}000{,}000{,}000 = 3.725 \times 10^{12}$

COMPETENCY 3.0   UNDERSTAND OPERATIONS ON REAL NUMBERS

Skill 3.1   **Identify, analyze, and apply number properties and number operations (e.g., commutative and associative properties, inverses, order of operations)**

Properties are rules that apply for addition, subtraction, multiplication, or division of real numbers. These properties are:

Commutative: The order of the terms or factors can be changed without changing the result.

Commutative Property of Addition: $a + b = b + a$

Example: $5 + (^-8) = {^-8} + 5 = {^-3}$

Commutative Property of Multiplication: $ab = ba$

Example: $^-2 \times 6 = 6 \times (^-2) = {^-12}$

The commutative property does not apply to subtraction or division because changing the order of the terms changes the result ($4 - 3 \neq 3 - 4$).

Associative: The terms can be regrouped without changing the result.

Associative Property of Addition: $a + (b + c) = (a + b) + c$

Example: $(^-2 + 7) + 5 = {^-2} + (7 + 5)$
$5 + 5 = {^-2} + 12 = 10$

Associative Property of Multiplication: $a(bc) = (ab)c$

Example: $(3 \times {^-7}) \times 5 = 3 \times ({^-7} \times 5)$
$^-21 \times 5 = 3 \times {^-35} = {^-105}$

This rule does not apply for division and subtraction.

Identity: The number is unchanged by the operation.

Additive Identity: The sum of any number and zero is that number.
$a + 0 = a$

Example: 17 + 0 = 17

Multiplicative Identity: The product of any number and one is that number.
$a \cdot 1 = a$

Example: ‾34 × 1 = ‾34

Inverse: The operation results in 0 or 1.

Additive Inverse: (-a) is the additive inverse of a therefore,
$a + (-a) = 0$

Example: 25 + (‾25) = 0

Multiplicative Inverse: (1/a), also called the reciprocal, is the multiplicative inverse of a. $a \cdot (1/a) = 1$

Example: $5 \times \dfrac{1}{5} = 1$

Distributive Property: The terms within parentheses can be multiplied by another term individually without first performing operations within the parentheses. This is useful when terms within the parentheses cannot be combined.

$a (b + c) = ab + ac$

Example: 6 × (‾4 + 9) = (6 × ‾4) + (6 × 9)
        6 × 5 = ‾24 + 54 = 30

To multiply a sum by a number, multiply each addend by the number, then add the products.

Example: Simplify 5(20 + 8).
100 + 40 = 140

The Order of Operations are to be followed when evaluating algebraic expressions. Follow these steps in order:

1. Simplify inside grouping characters such as parentheses, brackets, square root, fraction bar, etc.

2. Multiply out expressions with exponents.

3. Do multiplication or division, from left to right.

4. Do addition or subtraction, from left to right.

Example: Simplify. $3^3 - 5(b + 2)$

$$= 3^3 - 5b - 10$$

$$= 27 - 5b - 10 = 17 - 5b$$

Example: Simplify. $2 - 4 \times 2^3 - 2(4 - 2 \times 3)$

$$= 2 - 4 \times 2^3 - 2(4 - 6) = 2 - 4 \times 2^3 - 2(^-2)$$

$$= 2 - 4 \times 2^3 + 4 = 2 - 4 \times 8 + 4$$

$$= 2 - 32 + 4 = 6 - 32 = ^- 26$$

Skill 3.2   **Use properties of number operations to justify computational procedures**

See Skill 3.1

Skill 3.3   **Apply the laws of exponents to simplify expressions**

The **exponent form** is a method to write repeated multiplication. Basic form: b^n, where b is called the base and n is the exponent. b and n are both real numbers. b^n implies that the base b is multiplied by itself n times.

Example: Simplify $3^4 = 3 \times 3 \times 3 \times 3 = 81$

$$2^3 = 2 \times 2 \times 2 = 8$$

$$(^-2)^4 = (^-2) \times (^-2) \times (^-2) \times (^-2) = 16$$

$$^-2^4 = ^-(2 \times 2 \times 2 \times 2) = ^-16$$

Key exponent rules

For 'a' nonzero, and 'm' and 'n' real numbers:

1) $a^m \cdot a^n = a^{(m+n)}$ Product rule

2) $\dfrac{a^m}{a^n} = a^{(m-n)}$ Quotient rule

3) $\dfrac{a^{-m}}{a^{-n}} = \dfrac{a^n}{a^m}$

Example: Simplify $5^7 + 5^4$.
$5^7 + 5^4 = 5^{7+4} = 5^{11}$

Example: Simplify $\dfrac{4^6}{4^5}$.

$$\dfrac{4^6}{4^5} = 4^{6-5} = 4^1 = 4$$

When 10 is raised to any power, the exponent tells the numbers of zeroes in the product.

Example: Solve.

$$10^7 = 10,000,000$$

Caution: Unless the negative sign is inside the parentheses and the exponent is outside the parentheses, the sign is not affected by the exponent.

$(^-2)^4$ implies that -2 is multiplied by itself 4 times.

$^-2^4$ implies that 2 is multiplied by itself 4 times, then the answer is negated.

Skill 3.4 **Solve problems using real numbers in applied contexts**

Numbers are everywhere, at the gas station, in the weather forecast, in the ups and downs of the stock market. Shopping is the most common real-world situation in which mathematical skills are needed. Following are several examples of the application of mathematics to everyday activities.

To find the amount of sales tax on an item, change the percent of sales tax into an equivalent decimal number by moving the decimal point two places to the left. Then multiply the decimal number by the price of the object to find the sales tax. The total cost of an item will be the price of the item plus the sales tax.

Example: A guitar costs $120 plus 7% sales tax. How much are the sales tax and the total bill?

7% = 0.07 in decimal form
(0.07)($120) = $8.40 sales tax
$120 + $8.40 = $128.40 ← total cost

An alternative method to find the total cost is to multiply the price times the factor 1.07 (price + sales tax):

$$\$120 \times 1.07 = \$128.40$$

This gives you the total cost in fewer steps.

Example: A suit costs $450 plus 6½% sales tax. How much are the sales tax and the total bill?

> 6½% = 0.065 in decimal form
> (0.065)(450) = $29.25 sales tax
> $450 + $29.25 = $479.25 ← total cost

Using the alternative method to find total cost, multiply the price times the factor 1.065 (price + sales tax):

> $450 × 1.065 = $479.25

This gives you the total cost in fewer steps.

Another kind of real-world mathematical calculation involves time. Elapsed time problems are usually one of two types. One type of problem is the elapsed time between two times given in hours, minutes and seconds. The other common type of problem is between two times given in months and years.

For any time of day past noon, change to military time by adding 12 hours. For instance, 1:15 p.m. would be 13:15. Remember when you borrow a minute or an hour in a subtraction problem that you have borrowed 60 more seconds or minutes.

Example: Find the time from 11:34:22 a.m. until 3:28:40 p.m.

> Convert 3:28:40 p.m. to military time
>   3:28:40
> +12:00:00
>   15:28:40
> Now subtract
>   15:28:40
> −11:34:22
>   :18
> Borrow an hour and add 60 more minutes. Subtract.
>   14:88:40
> − 11:34:22
>   3:54:18   ↔ 3 hours, 54 minutes, 18 seconds

Example: John lived in Arizona from September 1991 until March 1995. How long is that?

		year	month
March 1995	=	95	03
September 1991	=	−91	09

Borrow a year and convert it into 12 more months, **subtract**.

		year	month
March 1995	=	94	15
September 1991	=	−91	09
		3 yrs	6 months

Example: A race took the winner 1 hr. 58 min. 12 sec. on the first half of the race and 2 hr. 9 min. 57 sec. on the second half of the race. How much time did the entire race take?

```
    1 hr.  58 min.  12 sec.
  + 2 hr.   9 min.  57 sec.  Add
    3 hr.  67 min.  69 sec.
         +  1 min. −60 sec.  Convert 60 sec. to 1 min.
    3 hr.  68 min.   9 sec.
  + 1 hr.−60 min.             Convert 60 min. to 1 hr.
    4 hr.   8 min.   9 sec. ← Final answer
```

SUBAREA II.   **MEASUREMENT AND GEOMETRY**

COMPETENCY 4.0   UNDERSTAND THE PRINCIPLES AND APPLICATIONS OF MEASUREMENT

Skill 4.1   Use appropriate units of measurement (e.g., length, area, volume, angles, weight, temperature, time, rates of change) to solve a variety of problems

A system of measurement is a set of units that can be used to express a measurement. There are two systems of measurement – Metric and Customary. The Metric system is based on powers of ten. The Customary system is primarily used in the U.S.

Measurements of length (Customary system)

12 inches (in)	=	1 foot (ft)
3 feet (ft)	=	1 yard (yd)
1,760 yards (yd)	=	1 mile (mi)

Measurements of length (Metric system)

kilometer (km)	=	1,000 meters (m)
hectometer (hm)	=	100 meters (m)
decameter (dam)	=	10 meters (m)
meter (m)	=	1 meter (m)
decimeter (dm)	=	1/10 meter (m)
centimeter (cm)	=	1/100 meter (m)
millimeter (mm)	=	1/1,000 meter (m)

Measurements of weight (Customary system)

28.35 grams (g)	=	1 ounce (oz)
16 ounces (oz)	=	1 pound (lb)
2,000 pounds (lb)	=	1 ton (t) (short ton)
1.1 ton (t)	=	1 metric ton (t)

Measurements of mass (Metric system)

1 kilogram (kg)	=	1,000 grams (g)
1 gram (g)	=	1,000 milligrams (mg)
1 milligram (mg)	=	1/1,000 gram (g)

Measurement of capacity (Customary system)

8 fluid ounces (fl oz)	=	1 cup (c)
2 cups (c)	=	1 pint (pt)
2 pints (pt)	=	1 quart (qt)
4 quarts (qt)	=	1 gallon (gal)

Measurement of capacity (Metric system)

1 kiloliter (kL)	=	1,000 liters (L)
1 liter (L)	=	1,000 milliliters (mL)
1 milliliter (mL)	=	1/1,000 liter (L)

Note: (') represents feet and (") represents inches.

Example: The distance around a race course is exactly 1 mile, 17 feet, and $9\frac{1}{4}$ inches. Approximate this distance to the nearest tenth of a foot.

Convert the distance to feet.

1 mile = 1,760 yards = 1,760 × 3 feet = 5,280 feet.

$$9\frac{1}{4} \text{ in.} = \frac{37}{4}\text{in.} \times \frac{1 \text{ ft}}{12 \text{ in.}} = \frac{37}{48} \text{ ft} = 0.77083 \text{ ft}$$

So 1 mile, 17 ft. and $9\frac{1}{4}$ in. = 5,280 ft + 17 ft + 0.77083 ft

= 5,297.$\underline{7}$7083 ft

Now, we need to round to the nearest tenth digit. The underlined 7 is in the tenths place. The digit in the hundredths place, also a 7, is greater than 5, so the 7 in the tenths place needs to be rounded up to 8 to get a final answer of 5,297.8 feet.

Example: Kathy has a bag of potatoes that weighs 5 lbs. 10 oz. She uses one third of the bag to make mashed potatoes. How much does the bag weigh now?

1 lb. = 16 oz

5(16 oz) + 10 oz = 80 oz+ 10 oz = 90 oz

$90 - (\frac{1}{3})90$ oz = 90 oz − 30 oz = 60 oz

60 ÷ 16 = 3.75 lbs

<u>Example:</u> The weight limit of a playground merry-go-round is 1,000 pounds. There are 11 children on the merry-go-round.
3 children weigh 100 pounds.
6 children weigh 75 pounds
2 children weigh 60 pounds

George weighs 80 pounds. Can he get on the merry-go-round?
3(100 lbs) + 6(75 lbs) + 2(60 lbs)
= 300 lbs + 450 lbs + 120 lbs
= 870 lbs
1000 lbs − 870 lbs
= 130 lbs

Since 80 lbs is less than 130 lbs, George can go on the merry-go-round.

<u>Example:</u> Students in a fourth grade class want to fill a 3-gallon jug using cups of water. How many cups of water are needed?

1 gallon = 16 cups of water

$$3 \text{ gal} \times \frac{16 \text{ c}}{1 \text{ gal}} = 48 \text{ c water are needed.}$$

Square units can be derived with knowledge of basic units of length by squaring the equivalent measurements.

1 square foot (sq. ft) = 144 sq. in.
1 sq. yd = 9 sq. ft
1 sq. yd = 1,296 sq. in.

Example: Solve.

14 sq. yd = _____ sq. ft
$14 \times 9 = 126$ sq. ft

Time

1 hour (h)	=	60 minutes (min)
1 minute (min)	=	60 seconds (s)
1 day	=	24 hours (h)
1 week	=	7 days

Example: It takes Cynthia 45 minutes to get ready each morning. How many hours does she spend getting ready each week?

$$\frac{45 \text{ min}}{1 \text{ day}} \times \frac{7 \text{ days}}{1 \text{ week}} = 315 \text{ min/week}$$

$315 \div 24 = 13.125$ h

For any time of day past noon, change to military time by adding 12 hours. For instance, 1:15 p.m. would be 13:15. Remember when you borrow a minute or an hour in a subtraction problem that you have borrowed 60 more seconds or minutes.

Example: Find the time from 11:34:22 a.m. until 3:28:40 p.m.

Convert 3:28:40 p.m. to military time
      3:28:40
   +12:00:00
    15:28:40
Now subtract
    15:28:40
  −11:34:22
          :18
Borrow an hour and add 60 more minutes. Subtract.
    14:88:40
  − 11:34:22
      3:54:18 ↔ 3 hours, 54 minutes, 18 seconds

Example: John lived in Arizona from September 1991 until March 1995. How long is that?

	year	month
March 1995 =	95	03
September 1991 =	−91	09

Borrow a year and convert it into 12 more months, **subtract**.

	year	month
March 1995 =	94	15
September 1991 =	−91	09
		3 yrs 6 months

Temperature

Temperature can be measured on three different scales – Celsius, Fahrenheit, and Kelvin. Water freezes at 0°C and 32°F. Water boils at 100°C and 212°F.

To convert a Celsius temperature to Fahrenheit, use the formula:

$$F = \frac{9}{5}C + 32$$

To convert a Fahrenheit temperature to Celsius, use the formula:

$$C = \frac{5}{9}(F - 32)$$

Example: The temperature one evening in Macon, GA, is 78°F. What is the temperature in degrees Celsius?

$$C = \frac{5}{9}(78 - 32)$$

$$= \frac{5}{9}(46)$$

$$= 25.6°C$$

Perimeter and area

The **perimeter** of any polygon is the sum of the lengths of the sides.

The **area** of a polygon is the number of square units covered by the figure or the space that a figure occupies.

FIGURE	AREA FORMULA	PERIMETER FORMULA
Rectangle	LW	$2(L + W)$
Triangle	$\frac{1}{2}bh$	$a + b + c$
Parallelogram	bh	sum of lengths of sides
Trapezoid	$\frac{1}{2}h(b_1 + b_2)$	sum of lengths of sides

<u>Example:</u> A farmer has a piece of land shaped as shown below. He wishes to fence this land at an estimated cost of $25 per linear foot. What is the total cost of fencing this property to the nearest dollar?

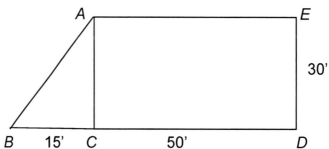

From the right triangle ABC, AC = 30 and BC = 15.

Since $(AB) = (AC)^2 + (BC)^2$
$(AB) = (30)^2 + (15)^2$

So $\sqrt{(AB)^2} = AB = \sqrt{1125} = 33.5410$ feet

Perimeter of the piece of land is $= AB + BC + CD + DE + EA$

= 33.5 + 15 + 50 + 30 + 50 = 178.5 feet
Cost of fencing = $25 x 178.5 = $4, 463.00

<u>Example:</u> What will be the cost of carpeting a rectangular office that measures 12 feet by 15 feet if the carpet costs $12.50 per square yard?

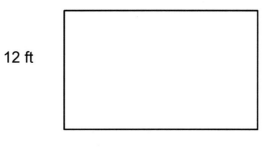

12 ft

15 ft

The problem is asking to determine the area of the office. The area of a rectangle is A = length x width.
Substitute the given values in the equation A = lw

   A = (12 ft)(15 ft)
   A = 180 ft^2

The problem asked to determine the cost of carpet at $12.50 per square yard.
Convert 180 ft^2 into yards^2.

   1 yd = 3 ft
   (1 yd)(1 yd) = (3 ft)(3 ft)
   1 yd^2 = 9 ft^2

$$180 \text{ ft}^2 \times \frac{1 \text{ yd}^2}{9 \text{ ft}^2} = 20 \text{ yd}^2$$

The carpet cost $12.50 per square yard; thus the cost of carpeting the office described is $12.50 x 20 = $250.00.

<u>Example:</u> Find the area of a parallelogram whose base is 6.5 cm and the height of the altitude to that base is 3.7 cm.

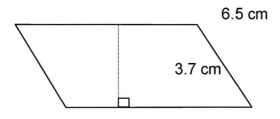

6.5 cm

3.7 cm

$A_{parallelogram}$ = bh
= (3.7 cm)(6.5 cm)
= 24.05 cm^2

Example: Find the area of the triangle below.

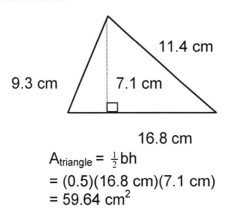

16.8 cm

$A_{triangle} = \frac{1}{2}bh$
$= (0.5)(16.8 \text{ cm})(7.1 \text{ cm})$
$= 59.64 \text{ cm}^2$

Example: Find the area of the trapezoid below.

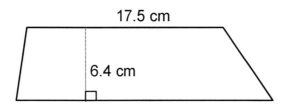

23.7 cm

The area of a trapezoid equals one-half the sum of the bases times the height.

$A_{trapezoid} = \frac{1}{2}h(b_1 + b_2)$
$= 0.5 \, (6.4 \text{ cm}) \, (17.5 \text{ cm} + 23.7 \text{ cm})$
$= 131.84 \text{ cm}^2$

The distance around a circle is the **circumference**. The ratio of the circumference to the diameter is represented by the Greek letter pi. $\pi \sim 3.14$.
The circumference of a circle is found by the formula $C = 2\pi r$ or $C = \pi d$ where r is the radius of the circle and d is the diameter.

The **area** of a circle is found by the formula $A = \pi r^2$.

<u>Example:</u> Find the circumference and area of a circle whose radius is 7 meters.

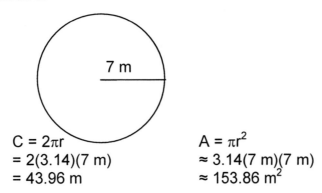

$C = 2\pi r$

$= 2(3.14)(7 \text{ m})$

$= 43.96 \text{ m}$

$A = \pi r^2$

$\approx 3.14(7 \text{ m})(7 \text{ m})$

$\approx 153.86 \text{ m}^2$

Volume and **Surface area** are computed using the following formulas:

FIGURE	VOLUME	SURFACE AREA
Right Cylinder	$\pi r^2 h$	$2\pi rh + 2\pi r^2$
Right Cone	$\dfrac{\pi r^2 h}{3}$	$\pi r\sqrt{r^2 + h^2} + \pi r^2$
Sphere	$\dfrac{4}{3}\pi r^3$	$4\pi r^2$
Rectangular Solid	LWH	$2LW + 2WH + 2LH$

FIGURE	LATERAL AREA	SURFACE AREA	VOLUME
Regular Pyramid	1/2Pl	1/2Pl+B	1/3Bh

P = Perimeter
h = height
B = Area of Base
l = slant height

<u>Example:</u> What is the volume of a shoebox with a length of 35 cm, a width of 20 cm and a height of 15 cm?

Volume of a rectangular solid = Length x Width x Height

$= (35 \text{ cm})(20 \text{ cm})(15 \text{ cm})$

$= 10,500 \text{ cm}^3$

Example: A water company is trying to decide whether to use traditional cylindrical paper cups or to offer conical paper cups since both cost the same. The traditional cups are 8 cm wide and 14 cm high. The conical cups are 12 cm wide and 19 cm high. The company will use the cup that holds the most water. Which cup will the company choose?

Draw and label a sketch of each.

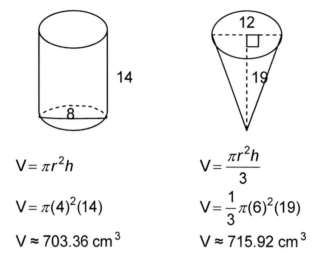

$V = \pi r^2 h$

$V = \pi(4)^2(14)$

$V \approx 703.36$ cm^3

$V = \dfrac{\pi r^2 h}{3}$

$V = \dfrac{1}{3}\pi(6)^2(19)$

$V \approx 715.92$ cm^3

The choice should be the conical cup since its volume is greater.

Example: How much material is needed to make a basketball that has a diameter of 15 inches? How much air is needed to fill the basketball?

Draw and label a sketch:

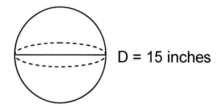

D = 15 inches

Surface Area

$SA = 4\pi r^2$

$= 4\pi(7.5)^2$

≈ 706.5 in.^{2}

Volume

$V = \dfrac{4}{3}\pi r^3$

$= \dfrac{4}{3}\pi(7.5)^3$

$\approx 1{,}766.25$ in.^{3}

Problems Involving Rates

Example: A class wants to take a field trip from Augusta to Atlanta to visit the capital. The trip is approximately 160 miles. If they will be traveling at 50 miles per hour, how long will it take for them to get there (assuming traveling at a steady rate)?

Set up the equation as a proportion and solve:

$$\frac{160 \text{ miles}}{x \text{ hours}} = \frac{50 \text{ miles}}{1 \text{ hour}}$$

(160 miles)(1 hour) = (50 miles)(x hours)

160 = 50x

x = 3.2 hours

Example: A salesperson drove 480 miles from Pittsburgh, PA, to Hartford, CT. The next day he returned the same distance to Pittsburgh in half an hour less time than his original trip took, because he increased his average speed by 4 mph. Find his original speed.

Since distance = rate x time then time = $\dfrac{\text{distance}}{\text{rate}}$

original time $-$ 1/2 hour $=$ shorter return time

$$\frac{480}{x} - \frac{1}{2} = \frac{480}{x+4}$$

Multiplying by the LCD of $2x(x+4)$, the equation becomes:

$$480\left[2(x+4)\right] - 1\left[x(x+4)\right] = 480(2x)$$

$$960x + 3840 - x^2 - 4x = 960x$$

$$x^2 + 4x - 3840 = 0$$

$(x+64)(x-60) = 0$ Either (x-60=0) or (x+64=0) or both=0

$x = 60$ 60 mph is the original speed.

$x = 64$ This is the solution since the time

cannot be negative. Check your answer

$$\frac{480}{60} - \frac{1}{2} = \frac{480}{64}$$

$$8 - \frac{1}{2} = 7\frac{1}{2}$$

$$7\frac{1}{2} = 7\frac{1}{2}$$

Cost per Unit

The unit rate for purchasing an item is its price divided by the number of units (pounds, ounces, etc.) of the item. The item with the lower unit rate is the lower price.

Example: Find the item with the best unit price:

$1.79 for 10 ounces
$1.89 for 12 ounces
$5.49 for 32 ounces

$$\frac{\$1.79}{10}=\$0.179/oz \qquad \frac{\$1.89}{12}=\$0.1575/oz \qquad \frac{\$5.49}{32}=\$0.172/oz$$

$1.89 for 12 ounces is the best price.

Skill 4.2 **Convert measurements within the customary and metric systems**

Conversions can be made to customary and metric units by multiplying by the appropriate conversion factor.

Length
1 inch	=	2.54 centimeters
1 foot	≈	30.48 centimeters
1 yard	≈	0.91 meters
1 mile	≈	1.61 kilometers

Example: A car skidded 170 yards on an icy road before coming to a stop. How long is the skid distance in kilometers?

Since 1 yard ≈ 0.9 meters, multiply 170 yards by 0.9 meters/1 yard.

$$170 \text{ yd} \times \frac{0.9 \text{ m}}{1 \text{ yd}}=153 \text{ m}$$

Since 1,000 meters = 1 kilometer, multiply 153 meters by 1 kilometer/1,000 meters.

$$153 \text{ m} \times \frac{1 \text{ km}}{1,000 \text{ m}}=0.153 \text{ km}$$

Weight and Mass

1 ounce	≈	28.35 grams
1 pound	≈	0.454 kilogram
1.1 ton	=	1 metric ton

Example: Zachary weighs 150 pounds. Tom weighs 153 pounds. What is the difference in their weights in grams?

153 lb – 150 lb = 3 lb

1 pound = 454 grams

$$3\text{ lb} \times \frac{454\text{ g}}{1\text{ lb}} = 1362\text{ g}$$

Capacity

1 teaspoon (tsp)	≈	5 milliliters
1 fluid ounce	≈	~~15~~ 29.57 milliliters
1 cup	≈	0.24 liters
1 pint	≈	0.47 liters
1 quart	≈	0.95 liters
1 gallon	≈	3.8 liters

Skill 4.3   **Demonstrate knowledge of strategies for estimating measurements and determining if answers are reasonable**

In order to estimate measurements, it is helpful to have a familiar reference with a known measurement. For instance, you can use the knowledge that a dollar bill is about six inches long or that a nickel weighs about 5 grams to make estimates of weight and length without actually measuring with a ruler or a balance.

Some common equivalents include:

ITEM	APPROXIMATELY EQUAL TO	
	METRIC	Customary
large paper clip	1 gram	1 ounce
capacity of sports bottle	1 liter	1 quart
average sized adult	75 kilograms	170 pounds
length of an office desk	1 meter	1 yard
math textbook	1 kilogram	2 pounds
length of dollar bill	15 centimeters	6 inches
thickness of a dime	1 millimeter	0.1 inches
area of football field		6,400 sq. yd
temperature of boiling water	100°C	212°F
temperature of ice	0°C	32°F
1 cup of liquid	240 mL	8 fl oz
1 teaspoon	5 ml	

Example: Estimate the measurement of the following items:

a) The length of an adult cow = _____meters
b) The thickness of a compact disc = _____millimeters
c) Your height = _____meters
d) length of your nose = _____centimeters
e) weight of your math textbook = _____kilograms
f) weight of an automobile = _____kilograms
g) weight of an aspirin = _____grams

a) 3

b) 2

c) 1.5

d) 4

e) 1

f) 1,000

g) 1

Depending on the degree of accuracy needed, an object may be measured to different units.

For example, a pencil may be 6 inches to the nearest inch, or 6-3/8 inches to the nearest eighth of an inch. Similarly, it might be 15 cm to the nearest cm or 154 mm to the nearest mm.

Skill 4.4 **Derive and apply formulas for perimeter, area, surface area, and volume for two and three-dimensional shapes and figures**

<u>Perimeter, area, and volume problems all require knowledge and understanding of formulas to calculate correct measurements. Students must identify which formula(s) are needed for a problem and correctly evaluate the formula(s).</u>

<u>Example:</u> A homeowner decided to fertilize her lawn. The shapes and dimensions of the lot, house, pool, and garden are given in the diagram below. The shaded area will not be fertilized. If each bag of fertilizer costs $7.95 and covers 4,500 square feet, find the total number of bags needed and the total cost of the fertilizer.

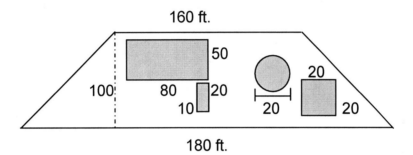

Area of Lot	Area of House	Area of Driveway
$A = \frac{1}{2} h(b_1 + b_2)$	$A = LW$	$A = LW$
$A = \frac{1}{2}(100)(180 + 160)$	$A = (80)(50)$	$A = (10)(25)$
$A = 17,000$ sq. ft.	$A = 4,000$ sq. ft.	$A = 250$ sq. ft.

Area of Pool	Area of Garden
$A = \pi r^2$	$A = s^2$
$A = \pi(10)^2$	$A = (20)^2$
$A = 314.159$ sq. ft.	$A = 400$ sq. ft.

Total area to fertilize = Lot area – (House + Driveway + Pool + Garden)

$$= 17,000 - (4,000 + 250 + 314.159 + 400)$$

$$= 12,035.841 \text{ sq ft}$$

Number of bags needed = Total area to fertilize / 4,500 sq.ft. bag

$$= 12,035.841 / 4,500$$

$$= 2.67 \text{ bags}$$

Since we cannot purchase 2.67 bags we must purchase 3 full bags.

Total cost = Number of bags * $7.95

$$= 3 * \$7.95$$

$$= \$23.85$$

Example: Find the total surface area of the given figure.

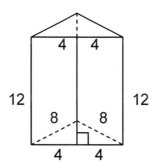

1. Since this is a triangular prism, first find the area of the bases.

2. Find the area of each rectangular lateral face.

3. Add the areas together.

$A = \dfrac{1}{2}bh$	$A = LW$	1. write formula
$8^2 = 4^2 + h^2$		2. find the height of
$h = 6.928$		the base triangle
$A = \dfrac{1}{2}(8)(6.928)$	$A = (8)(12)$	3. substitute values
$A = 27.713$ sq. units	$A = 96$ sq. units	4. compute

Total Area $= 2(27.713) + 3(96)$
$= 343.426$ sq. units

See Skill 4.1 for more examples.

Skill 4.5 **Use nets and cross sections to analyze three-dimensional figures**

Three-dimensional figures in geometry are called **solids**. A solid is the union of all points on a simple closed surface and all points in its interior. A **polyhedron** is a simple closed surface formed from planar polygonal regions. Each polygonal region is called a **face** of the polyhedron. The vertices and edges of the polygonal regions are called the **vertices** and **edges** of the polyhedron.

A **cylinder** is a solid has two congruent circular bases that are parallel.

A **sphere** is a solid having all its points the same distance from the center.

A **cone** is a solid having a circular base and a single vertex.

A **square pyramid** is a solid with a square base and 4 triangle-shaped faces.

A **tetrahedron** is a solid with 4 triangle faces.

A **prism** is a solid with two congruent, parallel bases that are polygons.

A **net** is a two-dimensional figure that can be cut out and folded up to make a three-dimensional solid. Below are examples of regular solids with their corresponding nets. Nets clearly show the shape and number of faces of a solid.

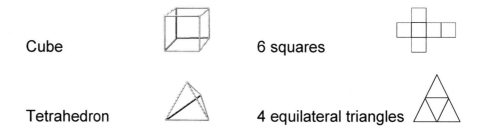

Cube   6 squares

Tetrahedron   4 equilateral triangles

<u>Example:</u> Draw the net of a triangular prism. Identify the polygons that make up the faces. How many vertices and edges do triangular prisms have?

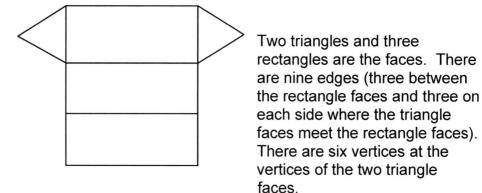

Two triangles and three rectangles are the faces. There are nine edges (three between the rectangle faces and three on each side where the triangle faces meet the rectangle faces). There are six vertices at the vertices of the two triangle faces.

Skill 4.6 **Apply the concepts of similarity, scale factors, and proportional reasoning to solve measurement problems**

Two figures that have the same shape are **similar**. Polygons are similar if and only if corresponding angles are congruent and corresponding sides are proportional.

Example: Given two similar quadrilaterals, find the lengths of sides *x, y,* and *z.*

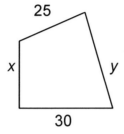
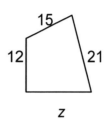

Since corresponding sides are proportional, 15/25 = 3/5, so the scale factor is 3/5.

$$\frac{12}{x} = \frac{3}{5} \qquad \frac{21}{y} = \frac{3}{5} \qquad \frac{z}{30} = \frac{3}{5}$$
$$3x = 60 \qquad 3y = 105 \qquad 5z = 90$$
$$x = 20 \qquad y = 35 \qquad z = 18$$

Example: Given the rectangles below, compare the area and perimeter.

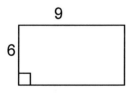
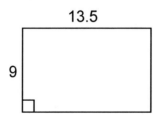

$A = LW$	$A = LW$	1. write formula
$A = (6)(9)$	$A = (9)(13.5)$	2. substitute known values
$A = 54$ sq. units	$A = 121.5$ sq. units	3. compute
$P = 2(L + W)$	$P = 2(L + W)$	1. write formula
$P = 2(6 + 9)$	$P = 2(9 + 13.5)$	2. substitute known values
$P = 30$ units	$P = 45$ units	3. compute

The areas relate to each other in the following manner:

Ratio of sides $9/13.5 = 2/3$

Multiply the area of the first polygon by the square of the reciprocal of the ratio, $(3/2)^2$, to get the area of the second.

$$54 \times (3/2)^2 = 121.5 \text{ sq. units}$$

The perimeters relate to each other in the following manner:

Ratio of sides $9/13.5 = 2/3$

Multiply the perimeter of the first by the reciprocal of the ratio to get the perimeter of the second.

$$30 \times 3/2 = 45 \text{ sq. units}$$

Example: Tommy draws and cuts out 2 triangles for a school project. The first triangle has sides that measure 3, 6, and 9 inches. The other triangle has corresponding sides measuring 2, 4, and 6 inches. Is there a relationship between the two triangles?

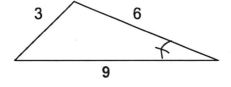 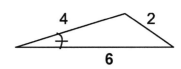

Determine the proportions of the corresponding sides.

$$\frac{2}{3} \qquad \frac{4}{6} = \frac{2}{3} \qquad \frac{6}{9} = \frac{2}{3}$$

The smaller triangle is 2/3 the size of the large triangle, therefore they are similar triangles.

COMPETENCY 5.0   UNDERSTAND THE PRINCIPLES AND APPLICATIONS OF EUCLIDEAN GEOMETRY

Skill 5.1   Analyze and apply properties of points, lines (e.g. parallel, perpendicular), planes, angles (e.g. complementary, corresponding), lengths, and distances (e.g. Pythagorean theorem)

A point, a line, and a plane are actually undefined terms since we cannot give a satisfactory definition using simple defined terms. However, their properties and characteristics give a clear understanding of what they are.

A **point** indicates a place or position in space. It has no length, width or thickness.

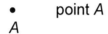

point A

A **line** is considered a straight set of points that does not end. Lines extend indefinitely in two directions.

A **plane** is a set of points composing a flat surface. A plane has no boundaries.

plane A

A **line segment** is a straight set of points that has two endpoints.

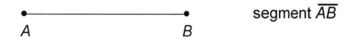

segment $\overline{AB}$

A **ray** has exactly one endpoint and extends indefinitely in one direction.

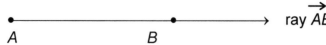

ray $\overrightarrow{AB}$

Perpendicular lines or planes form a 90° angle to each other. Perpendicular lines have slopes that are negative reciprocals of each other.

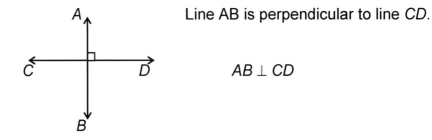

Line AB is perpendicular to line *CD*.

$AB \perp CD$

Parallel lines or planes do not intersect. Two parallel lines will have the same slope and are everywhere equidistant.

Line *AB* is parallel to line *CD*.

A ←——————————————→ B

C ←——————————————→ D

$AB \parallel CD$

Skew lines do not intersect because they do not lie on the same plane.

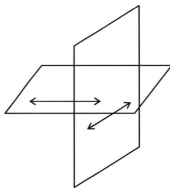

An **angle** is formed by the intersection of two rays.

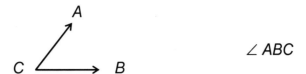

∠ ABC

Angles are measured in degrees. $1° = \frac{1}{360}$ of a circle.

A **right angle** measures 90°.

An **acute angle** measures more than 0° and less than 90°.

An **obtuse angle** measures more than 90° and less than 180°.

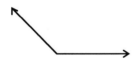

A **straight angle** measures 180°.

A **reflexive angle** measures more than 180° and less than 360°.

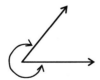

Angle relationships can be classified in a number of ways. Some classifications are outlined here.

Adjacent angles have a common vertex and one common side but no interior points in common.

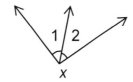

Complementary angles add up to 90°.

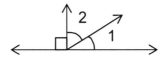

Supplementary angles add up to 180°.

Vertical angles have sides that form two pairs of opposite rays.

Corresponding angles are in the same corresponding position on two parallel lines cut by a transversal.

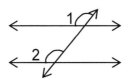

Parallel Lines Postulate: If two lines are parallel and are cut by a transversal, corresponding angles have the same measure.

Alternate interior angles are diagonal angles on the inside of two parallel lines cut by a transversal.

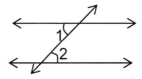

Alternate Interior Angles Theorem: If two parallel lines are cut by a transversal, the alternate interior angles are congruent.

Alternate exterior angles are diagonal angles on the outside of two parallel lines cut by a transversal. They are congruent.

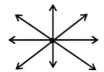

An infinite number of lines can be drawn through any point.

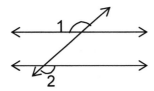

Exactly one line can be drawn through two points.

Like intersecting lines that share a common point, **intersecting planes** share a common set of points or line.

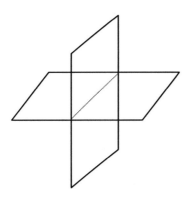

The Pythagorean Theorem states that given any triangle containing a right angle, $\triangle ABC$, the square of the hypotenuse is equal to the sum of the squares of the other two sides.

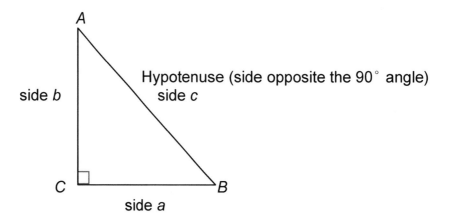

This theorem says that $(AB)^2 = (BC)^2 + (AC)^2$

or

$$c^2 = a^2 + b^2$$

Example: Find the area and perimeter of a rectangle if its length is 12 inches and its diagonal is 15 inches.

Draw and label sketch.

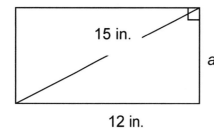

Use the Pythagorean Theorem

$$c^2 = a^2 + b^2$$
$$a^2 = 15^2 - 12^2$$
$$a^2 = 81$$
$$a = 9$$

Now use this information to find the area and perimeter.

$A = LW$ $P = 2(L + W)$
$A = (12)(9)$ $P = 2(12 + 9)$
$A = 108$ in.^{2} $P = 42$ in.

Skill 5.2 **Identify, analyze, and justify basic geometric constructions with compass and straightedge**

A geometric construction is a drawing made using only a compass and straightedge. A construction consists of only segments, arcs, and points. The easiest construction to make is to duplicate a given line segment. Given segment *AB*, construct a segment equal in length to segment *AB* by following these steps.

1. Place a point anywhere in the plane to anchor the duplicate segment. Call this point *S*.

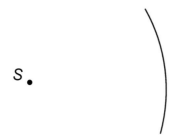

2. Open the compass to match the length of segment *AB*. Keeping the compass rigid, swing an arc from *S*.

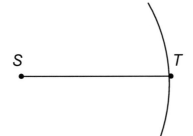

3. Draw a segment from *S* to any point on the arc. This segment will be the same length as *AB*.

Given a line such as line *l* and a point *P* not on *l*, follow these steps to construct a perpendicular line to l that passes through *P*.

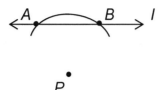

1. Swing an arc of any radius from *P* so that the arc intersects line *l* in two points *A* and *B*.

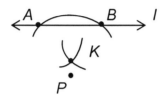

2. Open the compass to any length and swing two arcs of the same radius, one from *A* and the other from *B*. These two arcs will intersect at a new point *K*.

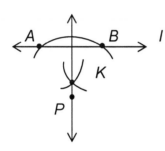

3. Connect *K* and *P* to form a line perpendicular to line *l* that passes through *P*.

Given a line segment with two endpoints such as *A* and *B*, follow these steps to construct the line that both bisects and is perpendicular to the line given segment.

1. Swing an arc of any radius from point *A*. Swing another arc of the same radius from *B*. The arcs will intersect at two points. Label these points *C* and *D*.

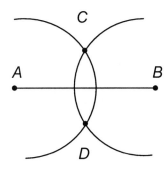

2. Connect *C* and *D* to form the perpendicular bisector of segment

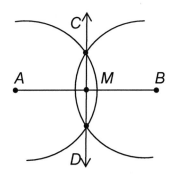

3. The point *M* where line *CD* and segment *AB* intersect is the midpoint of segment *AB*.

Skill 5.3   Analyze and apply the properties of similarity and congruence

Congruent figures have the same size and shape. If one is superimposed on the other, it will fit exactly. Congruent lines have the same length. Congruent angles have equal measures. The symbol for congruency is $\cong$.

Two triangles are congruent if each of the three angles and three sides of one triangle match up in a one-to-one fashion with congruent angles and sides of the second triangle. In order to see how the sides and angles match up, it is sometimes necessary to imagine rotating or reflecting one of the triangles so the two figures are oriented in the same position.

Following are some of the ways of proving that two triangles are congruent.

The **SSS Postulate** (side-side-side) states that if three sides of one triangle are congruent to the three corresponding sides of another triangle, then the two triangles are congruent.

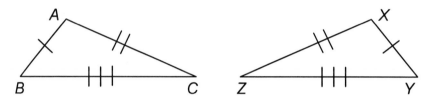

Since $AB \cong XY$, $BC \cong YZ$ and $AC \cong XZ$, $\triangle ABC \cong \triangle XYZ$.

<u>Example</u>: Given isosceles $\triangle ABC$ with D the midpoint of base AC, prove that $\triangle ABD$ is congruent to $\triangle CBD$.

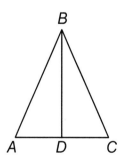

Proof:
1. Isosceles triangle *ABC*,
   D midpoint of base *AC*   Given
2. *AB* ≅ *BC*   An isosceles triangle has two
                                   congruent sides.
3. *AD* ≅ *DC*   The midpoint divides a segment
                                   into two equal parts.
4. *BD* ≅ *BD*   Reflexive Property
5. △*ABD* ≅ △*CBD*   SSS

The **SAS Postulate** (side-angle-side) states that if two sides and the included angle of one triangle are congruent to the corresponding two sides and the included angle of another triangle, then the two triangles are congruent.

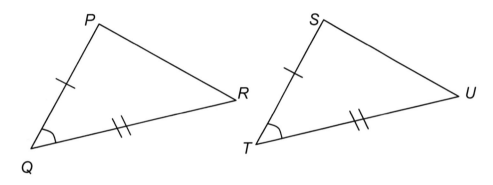

Since PQ ≅ ST, QR ≅ TU, and ∠Q ≅ ∠T, △PQR ≅ △STU

<u>Example</u>: Are the following triangles congruent?

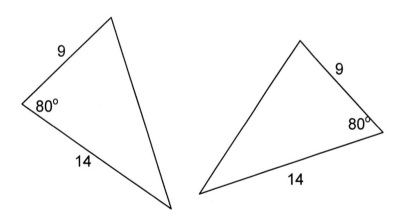

Each of the two triangles has a side that measures 14 units and another that measures 9 units. The angle included between the sides measures 80° in both triangles. Therefore the triangles are congruent by the SAS Postulate.

The **ASA Postulate** (angle-side-angle) states that if two angles and the included side of one triangle are congruent to the corresponding two angles and the included side of another triangle, the triangles are congruent.

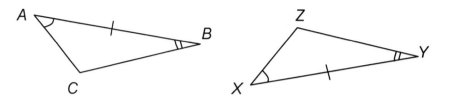

Since $\angle A \cong \angle X$, $\angle B \cong \angle Y$, and $AB \cong XY$, $\triangle ABC \cong \triangle XYZ$

<u>Example:</u> Given two right triangles with one leg (*AB* and *KL*) of each measuring 6 cm and the adjacent angle measuring 37°, prove the triangles are congruent.

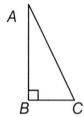

1. Right $\triangle ABC$ and $\triangle KLM$ $AB = KL = 6$ cm $\angle A = \angle K = 37°$	Given
2. $AB \cong KL$ $\angle A \cong \angle K$	Line segments and angles with the same measure are congruent
3. $\angle B \cong \angle L$	All right angles are congruent.
4. $\triangle ABC \cong \triangle KLM$	ASA

Example: What method could be used to prove the triangles are congruent?

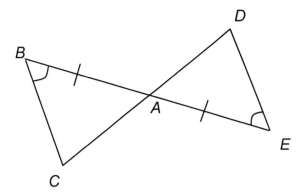

The sides *AB* and *AE* are given as congruent as are ∠*ABC* and ∠*DEA*. ∠*BAC* and ∠*DAE* are vertical angles and are therefore congruent. Thus △*ABC* and △*AED* are congruent triangles by the ASA Postulate.

The **AAS Theorem** (Angle- Angle-Side)states that if two angles of one triangle are congruent with the two corresponding angles in the other triangle and if the corresponding nonincluded sides are also congruent, then the two triangles are congruent.

The **HL Theorem** (Hypotenuse-Leg) is a congruence shortcut that may only be used with right triangles. According to this theorem, if the hypotenuse and leg of one right triangle are congruent to the hypotenuse and corresponding leg of the other right triangle, then the two triangles are congruent.

If two figures are congruent, then their corresponding sides and angles are also congruent.

Example: Polygons (pentagons) *ABCDE* and *VWXYZ* are congruent. They are exactly the same size and shape.

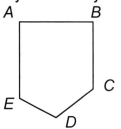 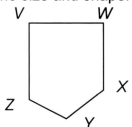

$$ABCDE \cong VWXYZ$$

Write the corresponding angles and sides of the two figures.

corresponding angles	corresponding sides
$\angle A \cong \angle V$	$AB \cong VW$
$\angle B \cong \angle W$	$BC \cong WX$
$\angle C \cong \angle X$	$CD \cong XY$
$\angle D \cong \angle Y$	$DE \cong YZ$
$\angle E \cong \angle Z$	$AE \cong VZ$

_Just as you can prove triangles congruent, you can also apply postulates and theorems to prove that they are similar.

The **Angle-Angle Similarity Postulate** (AA) states that if two triangles of one triangle are congruent to two angles of another triangle, then the triangles are similar.

The **Side-Angle-Side Similarity Theorem** (SAS) states that if an angle of one triangle is congruent to an angle of a second triangle, and the sides including the two angles are proportional, then the triangles are similar.

The **Side-Side-Side Similarity Theorem** (SSS) states that if the corresponding sides of two triangles are proportional, then the triangles are similar.

Example: If $\overline{AB}$ is parallel to $\overline{CD}$, prove that $\triangle KBA \sim \triangle KDC$.

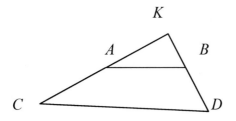

$\overline{AB} \parallel \overline{CD}$	Given
$\angle KAB \cong \angle KCD$	Corresponding angles are $\cong$.
$\angle CKD \cong \angle CKD$	Reflexive Property
$\triangle KBA \sim \triangle KDC$	AA Postulate

Skill 5.4 **Analyze and apply the properties of triangles, quadrilaterals, and other polygons to solve problems in applied contexts**

Polygons, simple, closed, **two-dimensional figures** composed of line segments, are named according to the number of sides they have.

A **triangle** is a polygon with three sides. Triangles can be classified by types of angles or lengths of sides they have. The sum of the measures of the angles of a triangle is 180°.

An **acute** triangle has three acute angles.
A **right** triangle has one right angle.
An **obtuse** triangle has one obtuse angle.

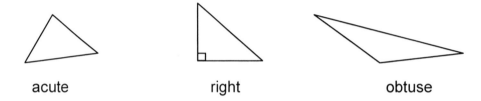

acute   right   obtuse

All three sides of an **equilateral** triangle are the same length.
Two sides of an **isosceles** triangle are the same length.
None of the sides of a **scalene** triangle are the same length.

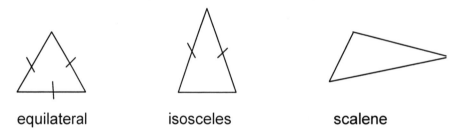

equilateral   isosceles   scalene

Example: Can a triangle have two right angles?

No. A right angle measures 90°, therefore the sum of two right angles would be 180° and there could not be third angle.

Example: Can a triangle have two obtuse angles?

No. Since an obtuse angle measures more than 90° the sum of two obtuse angles would be greater than 180°.

A **quadrilateral** is a polygon with four sides. The sum of the measures of the angles of a quadrilateral is 360°.

A **trapezoid** is a quadrilateral with one pair of parallel sides.

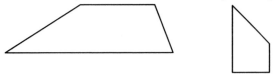

In an **isosceles trapezoid**, the non-parallel sides are congruent.

A **parallelogram** is a quadrilateral with two pairs of parallel sides. The diagonals bisect each other. Each diagonal divides the parallelogram into two congruent triangles. Both pairs of opposite sides are congruent. Both pairs of opposite angles are congruent. Two adjacent angles are supplementary.

A **rectangle** is a parallelogram with one right angle (and therefore all right angles).

A **rhombus** is a parallelogram with all sides of equal length.

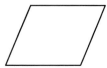

A **square** is a rectangle with all sides of equal length.

Example: True or false?

All squares are rhombuses.	True
All parallelograms are rectangles.	False - <u>some</u> parallelograms are rectangles
All rectangles are parallelograms.	True
Some rhombuses are squares.	True
Some rectangles are trapezoids.	False - only <u>one</u> pair of parallel sides
All quadrilaterals are parallelograms.	False - some quadrilaterals are parallelograms
Some squares are rectangles.	False - all squares are rectangles
Some parallelograms are rhombuses.	True

The sum of the measures of the **interior angles** of a polygon can be determined using the following formula, where *n* represents the number of angles in the polygon.

$$\text{Sum of } \angle s = 180(n - 2)$$

The measure of each angle of a regular polygon can be found by dividing the sum of the measures by the number of angles.

$$\text{Measure of } \angle = \frac{180(n - 2)}{n}$$

Example: Find the measure of each angle of a regular octagon. Since an octagon has eight sides, each angle equals:

$$\frac{180(8-2)}{8} = \frac{180(6)}{8} = 135°$$

The sum of the measures of the **exterior angles** of a polygon, taken one angle at each vertex, equals 360°.

The measure of each exterior angle of a regular polygon can be determined using the following formula, where n represents the number of angles in the polygon:

Measure of exterior $\angle$ of regular polygon

$$= 180 - \frac{180(n-2)}{n} \quad \text{or, more simply} \quad = \frac{360}{n}$$

Example: Find the measure of the interior and exterior angles of a regular pentagon.

Since a pentagon has five sides, each exterior angle measures:

$$\frac{360}{5} = 72°$$

Since each exterior angles is supplementary to its interior angle, the interior angle measures 180 − 72 or 108°.

Example: Two cars leave a road intersection at the same time. One car traveled due north at 55 mph while the other car traveled due east. After 3 hours, the cars were 180 miles apart. Find the speed of the second car.

Using a right triangle to represent the problem we get the figure:

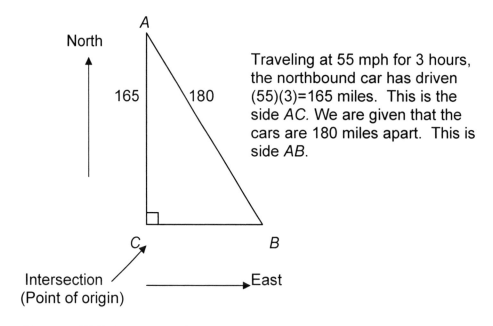

Traveling at 55 mph for 3 hours, the northbound car has driven (55)(3)=165 miles. This is the side AC. We are given that the cars are 180 miles apart. This is side AB.

Since △ABC is a right triangle, then, by Pythagorean Theorem, we get:

$$(AB)^2 = (BC)^2 + (AC)^2 \text{ or}$$
$$(BC)^2 = (AB)^2 - (AC)^2$$
$$(BC)^2 = 180^2 - 165^2$$
$$(BC)^2 = 32400 - 27225$$
$$(BC)^2 = 5175$$

Take the square root of both sides to get:
$$\sqrt{(BC)^2} = \sqrt{5175} \approx 71.937 \text{ miles}$$

Since the east bound car has traveled 71.935 miles in 3 hours, then the average speed is:

$$\frac{71.937}{3} \approx 23.97 \text{ mph}$$

COMPETENCY 6.0   UNDERSTAND COORDINATE AND TRANSFORMATIONAL GEOMETRY

Skill 6.1   Represent basic geometric figures in the coordinate plane

A Cartesian or rectangular **coordinate plane** consists of two perpendicular lines or axes that intersect at a point known as the origin. A unit of length is defined for each coordinate plane along with positive and negative directions for each axis. Traditionally, the horizontal axis is known as the *x*-axis (with positive direction to the right of the origin) and the vertical axis is the *y*-axis (with positive direction upward from the origin). The coordinate plane is divided into four quadrants by the intersection of the two axes as shown below.

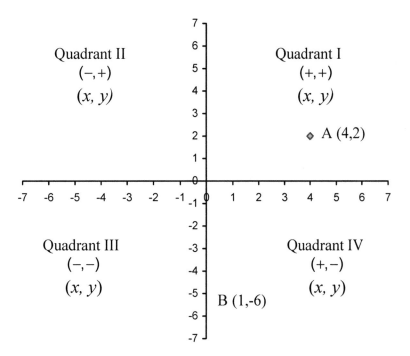

The **coordinates** of a point are a unique **ordered pair** of numbers that identify the location of the point on the coordinate plane. The first number in the ordered pair (x, y) identifies the position with regard to the x-axis while the second number identifies the position on the y-axis. Coordinate values of a point are determined by the distance of the point from the axes and the signs of the coordinates are determined by whether the point is in the positive or in the negative direction from the origin. In the coordinate plane shown above, point *A* is represented by the ordered pair (4, 2); point B is represented by the ordered pair (1, -6).

Geometric figures are represented as collections of points on the coordinate plane.

Example: Represent an isosceles triangle with two sides of length 4.

Draw the two sides along the *x*- and *y*- axes and connect the points (vertices). The vertices in this case are given by the points (0, 0), (4, 0) and (0, 4).

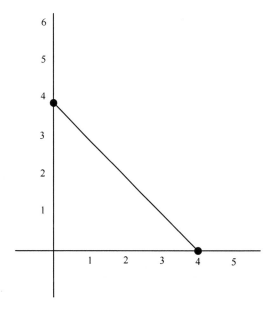

Skill 6.2 **Apply concepts and properties of slope, midpoint, parallelism, perpendicularity, distance, and congruence to solve problems in the coordinate plane and in applied contexts**

The **slope** of a straight line is the "slant" of the line. An upward left to right slant indicates a positive slope. A downward slant denotes a negative slope.

The formula for calculating the slope of a straight line that includes points (x_1, y_1) and (x_2, y_2) is:

$$\frac{y_2 - y_1}{x_2 - x_1}$$

The top of the fraction represents the change in the y coordinates; it is called the **rise**. The bottom of the fraction represents the change in the *x*-coordinates; it is called the **run,** and slope is often called "rise over run."

Example: Find the slope of a line that passes through the points (2, 2) and (7, 8).

$\dfrac{(8)-(2)}{(7)-(2)}$ plug the values into the formula

$\dfrac{6}{5}$ solve the rise over run

$= 1.2$ solve for the slope

The length of a line segment is the **distance** between the two endpoints (x_1, y_1) and (x_2, y_2). The distance formula is:

$$\sqrt{(x_1 - x_2)^2 + (y_1 - y_2)^2}$$

Example: Find the length between the points (2, 2) and (7, 8)

$= \sqrt{(2-7)^2 + (2-8)^2}$ plug the values into the formula

$= \sqrt{(-5)^2 + (-6)^2}$ calculate the x and y differences

$= \sqrt{25 + 36}$ square the values

$= \sqrt{61}$ add the two values

$= 7.81$ calculate the square root

Example: Find the perimeter of a figure with vertices at (4,5), ($^-4$, 6) and ($^-5$, $^-8$).

The figure being described is a triangle. Therefore, the length of each side must be found in order to evaluate the perimeter.

Side $1 = (4,5)$ to $(^-4,6)$ $D_1 = \sqrt{(^-4-4)^2 + (6-5)^2} = \sqrt{65}$

Side $2 = (^-4,6)$ to $(^-5, ^-8)$ $D_2 = \sqrt{(^-5-(^-4))^2 + (^-8-6)^2} = \sqrt{197}$

Side $3 = (^-5, ^-8)$ to $(4,5)$ $D_3 = \sqrt{((4-(^-5))^2 + (5-(^-8)^2))} = \sqrt{250}$

Thus, the perimeter of the figure $= \sqrt{65} + \sqrt{197} + 5\sqrt{10} \approx 37.9$ units

The **midpoint** of a line segment with endpoints (x_1, y_1) and (x_2, y_2), is given by:

$$\left(\dfrac{x_1 + x_2}{2}, \dfrac{y_1 + y_2}{2} \right)$$

<u>Example:</u> Find the center of a circle with a diameter whose endpoints are (3, 7) and ($^-$4, $^-$5).

$$\text{Midpoint} = \left(\frac{3 + (^-4)}{2}, \frac{7 + (^-5)}{2} \right)$$

$$= \left(\frac{^-1}{2}, 1 \right)$$

The slopes of different lines may be used to determine whether they are parallel or perpendicular to each other. **Parallel lines** have the same slope. **Perpendicular lines** have slopes that are negative reciprocals of each other.

<u>Example:</u> One line passes through the points (-4, -6) and (4, 6); another line passes through the points (-5, -4) and (3, 8). Are these lines parallel, perpendicular or neither?

Find the slopes.

$$m = \frac{y_2 - y_1}{x_2 - x_1}$$

$$m_1 = \frac{6 - (-6)}{4 - (-4)} = \frac{6 + 6}{4 + 4} = \frac{12}{8} = \frac{3}{2}$$

$$m_2 = \frac{8 - (-4)}{3 - (-5)} = \frac{8 + 4}{3 + 5} = \frac{12}{8} = \frac{3}{2}$$

Since the slopes are the same, the lines are parallel.

Example: One line passes through the points (1, -3) and (0, -6); another line passes through the points (4, 1) and (-2, 3). Are these lines parallel, perpendicular or neither?

Find the slopes.

$$m = \frac{y_2 - y_1}{x_2 - x_1}$$

$$m_1 = \frac{-6 - (-3)}{0 - 1} = \frac{-6 + 3}{-1} = \frac{-3}{-1} = 3$$

$$m_2 = \frac{3 - 1}{-2 - 4} = \frac{2}{-6} = -\frac{1}{3}$$

The slopes are negative reciprocals, so the lines are perpendicular.

Example: One line passes through the points (-2, 4) and (2, 5); another line passes through the points (-1, 0) and (5, 4). Are these lines parallel, perpendicular or neither?

Find the slopes.

$$m = \frac{y_2 - y_1}{x_2 - x_1}$$

$$m_1 = \frac{5 - 4}{2 - (-2)} = \frac{1}{2 + 2} = \frac{1}{4}$$

$$m_2 = \frac{4 - 0}{5 - (-1)} = \frac{4}{5 + 1} = \frac{4}{6} = \frac{2}{3}$$

Since the slopes are not the same, the lines are not parallel. Since they are not negative reciprocals, they are not perpendicular either. Therefore, the answer is "neither."

Skill 6.3 **Identify and apply concepts of symmetry and transformations (e.g., translations, rotations, reflections) to figures in the coordinate plane**

A **transformation** is a change in the position, shape, or size of a geometric figure. **Transformational geometry** is the study of manipulating objects by flipping, twisting, turning and scaling.

Symmetry exists if a figure is reflected or flipped across a line and appears unchanged. The line where the reflection occurs is known as the **line of symmetry**.

Example: Draw all the lines of symmetry in the figures below.

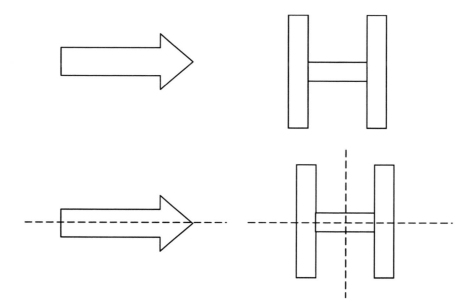

The four basic geometric transformations are **translation, rotation, reflection,** and **glide reflection**. The transformation of an object is called its image. If the original object was labeled with letters, such as *ABCD*, the image may be labeled with the same letters followed by a prime symbol, *A'B'C'D'* .

A **translation** is a transformation that "slides" an object a fixed distance in a given direction. The original object and its translation have the same shape and size, and they face in the same direction.

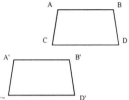

An example of a translation in architecture would be stadium seating. The seats are the same size and the same shape and face in the same direction. In the coordinate plane, a translation is achieved by changing the x and/or y coordinate for each point on the figure by a fixed number.

Example: What are the coordinates for a triangle defined by the vertices *A* (1, 2), *B* (6, 3) and *C*(-4, 5) and translated 2 units up?

A' (1, 4), *B'* (6, 5) and *C'* (-4, 7).

An object and its **reflection** have the same shape and size, but the figures face in opposite directions.

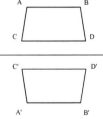

The line (where a mirror may be placed) is called the **line of reflection**. The distance from a point to the line of reflection is the same as the distance from the point's image to the line of reflection. In terms of coordinates, reflection through the x-axis, for instance, will reverse the sign of the y-coordinate of a point.

Example: What are the coordinates of a triangle defined by the vertices *D* (1, 2), *E* (6, 3) and *F* (-4 ,5) after a reflection about the x-axis?

D' (1, -2), *E'* (6, -3) and *F'* (-4, -5).

A **rotation** is a transformation that turns a figure about a fixed point called the center of rotation. An object and its rotation are the same shape and size, but the figures may be turned in different directions. Rotations can occur in either a clockwise or a counterclockwise direction.

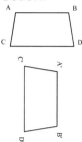

Rotations can be seen in wallpaper and art, and a Ferris wheel is an example of rotation.

Example: $\triangle XYZ$ has vertices at X (3, 4), Y (-6, -2) and Z (-1, 5). What are the coordinates of the vertices after the triangle has been rotated 90° counterclockwise about the origin?

X' (3, -4), Y' (-6, 2) and Z' (-1, -5).

A **glide reflection** is a combination of a reflection and a translation.

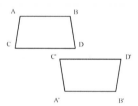

Many types of flooring found in our homes are examples of **symmetry**: Oriental carpets, tiling, patterned carpet, etc. The human body is an example of bilateral symmetry, even though it is not usually perfect. If you split the torso down the middle, on each half, you will find one ear, one eye, one nostril, one shoulder, one arm, one leg, and so on, in approximately the same place.

Skill 6.4   Use dilations and proportionality to analyze similar figures in the coordinate plane

Another type of transformation is the **dilation**. A dilation is a transformation that reduces or enlarges a figure by a scale factor. A dilation does not change the shape of the figure or the direction the figure is facing.

Example: Use a dilation to transform a figure.

Starting with a triangle whose center of dilation is point P,

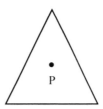

we dilate the lengths of the sides by the same factor, known as scale factor, to create a new triangle.

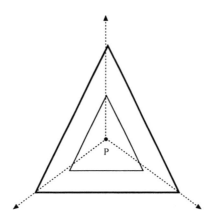

In terms of coordinates, dilation is implemented by multiplying each of the coordinates of the figure by the same factor.

Similar figures are related by dilation.

Example: $\triangle RST$ has vertices at R (0, 6), S (-5, 4) and T (2, -7). A second triangle, $\triangle IJK$, has vertices at I (0, 3), J $(-\frac{5}{2}, 2)$ and K (1, $-\frac{7}{2}$). Are the triangles similar? Explain your answer.

Yes, the triangles are similar. $\triangle IJK$ is a dilation of $\triangle RST$ with a scale factor of $\frac{1}{2}$. The measures of the sides of the triangles are proportionate to each other.

Skill 6.5 **Identify and describe three-dimensional figures formed by transformations of plane figures through space**

In order to represent three-dimensional figures, we need three coordinate axes (X, Y, and Z) which are all mutually perpendicular to each other. Since we cannot draw three mutually perpendicular axes on a two-dimensional surface, we use oblique representations.

Example: Represent a cube with sides of 2.

We draw three sides along the three axes to keep things simple.

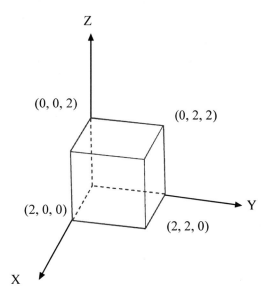

Each point has three coordinates (x, y, z).

For more information on solid figures see Skill 4.5.

SUBAREA III.   PATTERNS, ALGEBRA, AND FUNCTIONS

COMPETENCY 7.0   UNDERSTAND PATTERNS, RELATIONS, AND
                 FUNCTIONS

Skill 7.1   Analyze, extend, and describe a variety of patterns (e.g.,
               numerical, pictorial, recursive) using rules and algebraic
               expressions

Numerical patterns
A numerical pattern is a sequence of numbers arranged in a
particular order. Thus, given part of the sequence, one can use a
prescribed rule to find the numbers that follow or precede that part.
For instance, 1, 4, 9, 16...is a series that consists of the squares of
the natural numbers. Using this rule, the next term in the series, 25,
can be found by squaring the next natural number, 5.

Example: The following table represents the number of problems
Mr. Rodgers is assigning his math students for homework each
day, starting with the first day of class.

Day	1	2	3	4	5	6	7	8	9	10	11

Number of Problems	1	1	2	3	5	8	13

If Mr. Rodgers continues this pattern, how many problems will he
assign on the eleventh day?

If we look for a pattern, it appears that the number of problems
assigned each day is equal to the sum of the problems assigned for
the previous two days. We test this as follows:

          Day 2 = 1 + 0 = 1
          Day 3 = 1 + 1 = 2
          Day 4 = 2 + 1 = 3
          Day 5 = 3 + 2 = 5
          Day 6 = 5 + 3 = 8
          Day 7 = 8 + 5 = 13

Therefore, Day 8 would have 21 problems; Day 9, 34 problems;
Day 10, 55 problems; and Day 11, 89 problems.

A pattern may be expressed in terms of variables as in the example
below.

MIDDLE SCHOOL MATH.   78

Example: The equation $y = 2x + 1$ describes a pattern of points that all lie on the same straight line. A table of values is constructed in order to represent the pattern graphically.

$$
\begin{array}{cc}
x & y \\
-2 & -3 \\
-1 & -1 \\
0 & 1 \\
1 & 3 \\
2 & 5 \\
\end{array}
$$

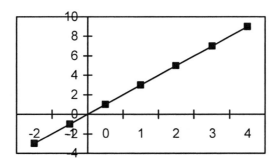

New points on this line may be found by picking an x-value and finding the corresponding y-value using the describing equation.

Arithmetic Sequences

An **arithmetic sequence** is a set of numbers with a common difference between the terms. Terms and the distance between terms can be calculated using the following formula:

$a_n = a_1 + (n-1)d$ where

    a_1 = the first term
    a_n = the nth term (general term)
    n = the number of the term in the sequence
    d = the common difference

Example: Find the 8th term of the arithmetic sequence 5, 8, 11, 14...

$a_n = a_1 + (n-1)d$
$a_1 = 5$ identify the 1st term
$d = 8 - 5 = 3$ find d
$a_8 = 5 + (8-1)3$ substitute
$a_8 = 26$

<u>Example:</u> Given two terms of an arithmetic sequence, find a_1 and d

$a_4 = 21$ $\qquad$ $a_6 = 32$

$a_n = a_1 + (n-1)d$ $\qquad$ $a_4 = 21, n = 4$

$21 = a_1 + (4-1)d$ $\qquad$ $a_6 = 32, n = 6$

$32 = a_1 + (6-1)d$

$21 = a_1 + 3d$ $\qquad$ solve the system of equations

$32 = a_1 + 5d$

$\begin{aligned} 32 &= a_1 + 5d \\ -21 &= -a_1 - 3d \\ \hline 11 &= 2d \end{aligned}$ $\qquad$ multiply by -1
add the equations

$5.5 = d$

$21 = a_1 + 3(5.5)$ $\qquad$ substitute $d = 5.5$ into either equation

$21 = a_1 + 16.5$

$a_1 = 4.5$

The sequence begins with 4.5 and has a common difference of 5.5 between numbers.

Geometric Sequences

A **geometric sequence** is a series of numbers in which a common ratio can be multiplied by a term to yield the next term. The common ratio can be calculated using the formula:

$r = \dfrac{a_{n+1}}{a_n}$ $\qquad$ where

r = common ratio

a_n = the nth term

The ratio is then used in the geometric sequence formula:

$a_n = a_1 r^{n-1}$

<u>Example:</u> Find the 8th term of the geometric sequence 2, 8, 32, 128...

$r = \dfrac{a_{n+1}}{a_n}$ $\qquad$ use common ratio formula to find ratio

$r = \dfrac{8}{2}$ $\qquad$ substitute $a_n = 2$, $a_{n+1} = 8$

$r = 4$

$a_n = a_1 \bullet r^{n-1}$ $\qquad$ use $r = 4$ to solve for the 8th term

$a_n = 2 \bullet 4^{8-1}$

$a_n = 32{,}768$

Just like the arithmetic and geometric sequences discussed before, other patterns can be created using algebraic variables. Patterns may also be pictorial. In each case, one can predict subsequent terms or find a missing term by first discovering the rule that governs the pattern.

Example: Find the next term in the sequence $ax^2y, ax^4y^2, ax^6y^3, \dots$

Inspecting the pattern we see that this is a geometric sequence with common ratio x^2y.

Thus, the next term $= ax^6y^3 \times x^2y = ax^8y^4$.

Example: Find the next term in the pattern:

Inspecting the pattern one observes that it has alternating squares and circles that include a number of hearts that increases by two for each subsequent term.

Hence, the next term in the pattern will be as follows:

Skill 7.2   Interpret and compare verbal and graphical descriptions of real-world relations

Number relations can be found in a variety of real-world situations. A **relation** is any set of ordered pairs.

For example, the Drama Club is washing cars for a fundraiser. The club earns $10 for each car the group washes. Thus, the drama club will earn $10 after washing one car, $20 after washing two cars, $30 after washing three cars, etc.

The relation can also be represented graphically.

Money Earned at Car Wash

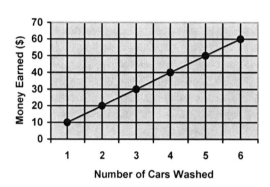

Example: The cost of making a call on a particular cell phone is $0.15 per minute. Describe and graph this relation. How much will it cost to make a 5-minute phone call?

The relation states that for every minute someone talks on this cell phone, he or she will be charged $0.15. So, a 5-minute call will cost $0.75. The relation is shown in the graph below.

Cell Phone Cost per Min

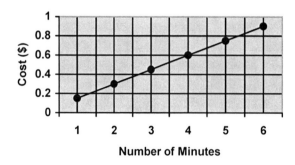

Skill 7.3   **Identify, describe, and analyze the properties of relations and functions (e.g., domain, range) in a variety of forms (e.g., tabular, graphical, algebraic)**

A **relation** is any set of ordered pairs.

The **domain** of a relation is the set made of all the first coordinates of the ordered pairs.

The **range** of a relation is the set made of all the second coordinates of the ordered pairs.

A **function** is a relation in which each value for the domain has a unique value for the range. (No x values are repeated.) On a graph, use the **vertical line test** to look for a function. If any vertical line intersects the graph of a relation in more than one point, then the relation is not a function.

A **mapping** is a diagram with arrows drawn from each element of the domain to the corresponding elements of the range. If two arrows are drawn from the same element of the domain, then it is not a function.

Example: Determine the domain and range of this mapping. Is the relation a function?

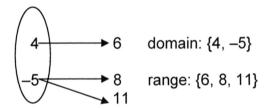

domain: {4, –5}

range: {6, 8, 11}

The relation is not a function because there are two values for –5.

Example: Are the mappings shown below true functions?

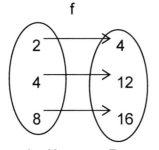

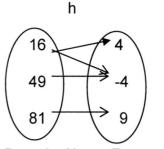

Domain, X   Range, Y   Domain, X   Range, Y

This is a "true" function.   This is not a "true" function.

A relation may also be described algebraically. An equation such as $y = 3x + 5$ describes a relation between the independent variable x and the dependent variable y. Thus, y is written as $f(x)$ "function of x."

Example: Given a function $f(x) = 3x + 5$, find $f(2)$; $f(0)$; $f(^-10)$

$f(2)$ denotes the value of the function f(x) at $x = 2$.

$$f(2) = 3(2) + 5 = 6 + 5 = 11$$
$$f(0) = 3(0) + 5 = 0 + 5 = 5 \qquad \text{Substitute for } x.$$
$$f(^-10) = 3(^-10) + 5 = ^-30 + 5 = ^-25$$

Example: Given $h(t) = 3t^2 + t - 9$, find $h(^-4)$.

$$h(^-4) = 3(^-4)^2 - 4 - 9$$
$$h(^-4) = 3(16) - 13$$
$$h(^-4) = 48 - 13 \qquad \text{Substitute for } t.$$
$$h(^-4) = 35$$

Relations can also be shown as tables and graphs. For example, the table below shows the ages and heights of six students in a class.

Age of Student	Height of Student (ft)
13	5.2
14	5.7
12	5.4
13	4.9
12	5.1
11	4.7

The relation is not a function because there are two different dependent values (5.2 ft and 4.9 ft) for the same independent value (13). There are also two different dependent values (5.4 ft and 5.1 ft) for the independent value of 12. The relation can be shown graphically:

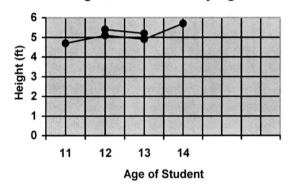

Heights of Students by Age

This relation does not pass the vertical-line test because if a vertical line was drawn down the 12- and 13-year ages, it would intersect 2 points instead of one.

See also Skill 9.2.

Skill 7.4 **Describe and analyze direct and inverse proportional relationships graphically and algebraically**

Two or more quantities can vary directly or inversely with each other. To express the relationship as an equation, we introduce another constant quantity called the constant of proportionality.

If y varies directly as x, then $y = kx$, where k is the constant of proportionality. k is the slope of the straight line that represents this relationship graphically (see example below).

If y varies inversely as x, then $y = \dfrac{k}{x}$ where k is the constant of proportionality. The graph of the relationship is not a straight line in this case.

If two variables vary directly, as one gets larger, the other also gets larger. If one gets smaller, then the other gets smaller too. If two variables vary inversely, as one gets larger, the other one gets smaller.

Example: If $30 is paid for 5 hours of work, how much would be paid for 19 hours of work?

Since pay varies directly with time working, we can write
$30 = c(5 \text{ hours})$ where c is the constant of proportionality.
Solving for c, which is the hourly pay rate in this case,
we find that c = $6/hour.
Thus, pay for 19 hours of work = ($6/hour)(19 hours) = $114.
This could also be represented

as a proportion: $\dfrac{\$30}{5} = \dfrac{y}{19}$

$$5y = \$570$$
$$y = \$114$$

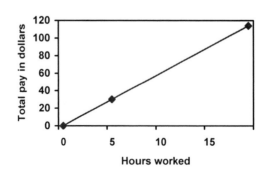

Example: On a 546-mile trip from Miami to Charlotte, one car drove 65 mph while another car drove 70 mph. How does this affect the driving time for the trip?

This is an inverse variation, since increasing your speed should decrease your driving time. Use the equation

$t = \dfrac{d}{r}$ where t=driving time, r = speed and d = distance traveled.

(65 mph)t = 546 miles   and   (70 mph)t = 546 miles
t = 8.4 hours   and   t = 7.8 hours
slower speed, more time   faster speed, less time

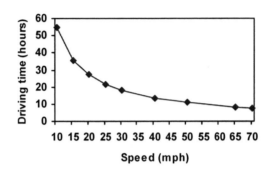

A quantity may also vary with different exponents of another quantity as shown in the examples below.

Example: A varies inversely as the square of R. When A = 2, R = 4. Find A if R = 10.

Since A varies inversely as the square of R,

$$A = \frac{k}{R^2} \text{ (equation 1)}, k \text{ is a constant.}$$

Use equation 1 to find k when A = 2 and R = 4.

$$2 = \frac{k}{4^2} \rightarrow 2 = \frac{k}{16} \rightarrow k = 32 .$$

Substituting k = 32 into equation 1 with R =10, we get:
$$A = \frac{32}{10^2} \rightarrow A = \frac{32}{100} \rightarrow A = 0.32$$

Example: x varies directly as the cube root of y and inversely as the square of z. When $x = 2$, $y = 27$ and $z = 1$. Find y when $z = 2$ and $x = 1$.

Since x varies directly as the cube root of y and inversely as the square of z.

$$x = k \cdot \frac{\sqrt[3]{y}}{z^2} \quad \text{(equation 1), } k \text{ is constant.}$$

Substituting in equation 1 to solve for k when $x = 2$, $y = 27$, and $z = 1$ we get:

$$2 = k \cdot \frac{\sqrt[3]{27}}{1^2} \rightarrow 2 = k \cdot \frac{3}{1} \rightarrow 2 = 3k$$

$$k = \frac{2}{3}$$

To solve for y when $z = 2$ and $x = 1$, we substitute in equation 1 using the value we found for k to get:

$$1 = \frac{2}{3} \cdot \frac{\sqrt[3]{y}}{2^2} \rightarrow 1 = \frac{2 \cdot \sqrt[3]{y}}{3(4)} \rightarrow 1 = \frac{2 \cdot \sqrt[3]{y}}{12}$$

$$1 = \frac{\sqrt[3]{y}}{6} \rightarrow 6 = \sqrt[3]{y} \quad \text{Cube both sides.}$$

$$6^3 = \left(\sqrt[3]{y}\right)^3$$

$$y = 216$$

Skill 7.5 **Describe and use various representations of nonlinear functions (e.g., quadratic, exponential)**

Nonlinear functions are those that are not in the form $y = ax + b$ (where x is the independent variable) and are not represented graphically by a straight line. These include functions with exponents of the independent variable that are not equal to one (e.g. $x^2, x^3, \frac{1}{x}$) and other special functions. A few of these are described below.

Quadratic functions

A quadratic function is written in the standard form $y = ax^2 + bx + c$ where the greatest exponent is 2.

The graphs of quadratic functions are parabolas. Parabolas are u-shaped curves that may open upward or downward and vary in width and steepness. To graph quadratic functions, it is best to first convert the function to standard form if it is not already in that form.

In standard form, we can tell if the graph opens upward (if a is positive) or downward (if a is negative). The smaller the value of $|a|$, the wider the parabola. We are also given the y-intercept of the parabola, which is represented by c.

To graph the parabola of a quadratic function, first make a table of values and choose at least five values for x. Substitute each value into the function and solve for y. Use the x- and y-values as coordinates and plot the ordered pairs. Continue to find and plot ordered pairs until the u-shaped curve is drawn.

Example: Graph the quadratic function $y = 3x^2 + x - 2$

After making a table of values, we get:

x	$y = 3x^2 + x - 2$	y
–2		8
–1		0
0		–2
1		2
2		12

We can now plot the ordered pairs (-2, 8), (-1, 0), (0, -2), (1, 2), and (2, 12):

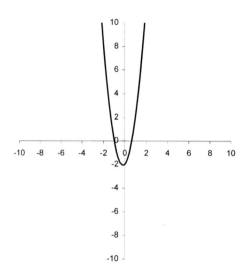

You can also graph a quadratic function by converting the function to vertex form. Vertex form for quadratic functions is $f(x) = a(x - h)^2 + k$. Working from vertex form, the vertical axis of symmetry is the line $x = h$. The vertex is the point (h, k). Finally, the parabola opens up if a is positive and down if a is negative.

Example: Graph the quadratic function $y = 3x^2 + x - 2$

Expressing this function in vertex form we get

$$y = 3(x + \frac{1}{6})^2 - \frac{25}{12}$$

Thus, the graph is a parabola with an axis of symmetry $x = -\frac{1}{6}$

and the vertex located at the point $(-\frac{1}{6}, -\frac{25}{12})$.

The zeros of a quadratic function may be found by solving the corresponding quadratic equation, i.e. the equation obtained by setting the function to zero. There are several different methods for solving quadratic equations.

One method of solving a quadratic equation is by **factoring** the quadratic expression and applying the condition that at least one of the factors must equal zero in order for the whole expression to be zero.

Example: Solve the equation $x^2 + 10x - 24 = 0$

$$x^2 + 10x - 24 = 0$$
$$(x + 12)(x - 2) = 0 \qquad \text{Factor.}$$
$$x + 12 = 0 \text{ or } x - 2 = 0 \qquad \text{Set each factor equal to 0.}$$
$$x = -12 \qquad x = 2 \qquad \text{Solve.}$$

Check:

$$x^2 + 10x - 24 = 0$$

$$(-12)^2 + 10(-12) - 24 = 0 \qquad (2)^2 + 10(2) - 24 = 0$$
$$144 - 120 - 24 = 0 \qquad 4 + 20 - 24 = 0$$
$$0 = 0 \qquad 0 = 0$$

A quadratic equation can also be solved by **completing the square**. In order to complete the square, the coefficient of the x^2 term must be 1. Isolate the x^2 and x terms. Then add the half of the coefficient of the x term squared to both sides of the equation. Finally take the square root of both sides and solve for x.

Example: Solve the equation $x^2 - 6x + 8 = 0$

$$x^2 - 6x + 8 = 0$$

$$x^2 - 6x = {}^-8 \qquad \text{Move the constant to the right side.}$$

$$x^2 - 6x + 9 = {}^-8 + 9 \qquad \text{Add the square of half the coefficient of } x \text{ to both sides.}$$

$$(x - 3)^2 = 1 \qquad \text{Write the left side as a perfect square.}$$

$$x - 3 = \pm\sqrt{1} \qquad \text{Take the square root of both sides.}$$

$$x - 3 = 1 \qquad x - 3 = {}^-1 \qquad \text{Solve.}$$
$$x = 4 \qquad x = 2$$

Check:

$$x^2 - 6x + 8 = 0$$

$$4^2 - 6(4) + 8 = 0 \qquad\qquad 2^2 - 6(2) + 8 = 0$$

$$16 - 24 + 8 = 0 \qquad\qquad 4 - 12 + 8 = 0$$

$$0 = 0 \qquad\qquad\qquad\qquad 0 = 0$$

To solve a quadratic equation using the **quadratic formula**, make sure your equation is in the form $ax^2 + bx + c = 0$.

The solution for x can be obtained by substituting the values of a, b, and c into the formula:

$$x = \frac{-b \pm \sqrt{b^2 - 4ac}}{2a}$$

Simplify the result to find the answers.

The discriminant of a quadratic equation is the part of the quadratic formula that is usually inside the radical sign, $b^2 - 4ac$.

$$x = \frac{-b \pm \sqrt{b^2 - 4ac}}{2a}$$

Note that the radical sign is **not** part of the discriminant.

If the value of the discriminant is **any negative number**, then there are **two complex roots** including "i."

If the value of the discriminant is **zero**, then there is only **one real rational root**. This would be a double root.

If the value of the discriminant is **any positive number that is also a perfect square**, then there are **two real rational roots**. (There are no longer any radical signs.)

If the value of the discriminant is **any positive number that is not a perfect square**, then there are **two real irrational roots**. (There are still unsimplified radical signs.)

Example: Find the value of the discriminant for the equation $2x^2 - 5x + 6 = 0$. Then determine the number and nature of the solutions of that quadratic equation.

$a = 2$, $b = {}^-5$, $c = 6$ so $b^2 - 4ac = ({}^-5)^2 - 4(2)(6) = 25 - 48 = {}^-23$.

Since $^-23$ is a negative number, there are **two complex roots** including "i".

$$x = \frac{5}{4} + \frac{i\sqrt{23}}{4}, \quad x = \frac{5}{4} - \frac{i\sqrt{23}}{4}$$

Some word problems can be solved by setting up a quadratic equation or inequality. Examples of this type of problem follow.

Example: The square of a number is equal to 6 more than the original number. Find the original number.

If x is the original number, we can write:

$x^2 = 6 + x$ Set this equal to zero.

$x^2 - x - 6 = 0$ Now factor.

$(x - 3)(x + 2) = 0$

$x = 3$ or $x = -2$ There are two solutions, 3 or -2.

Exponential Functions
Exponential functions of base a take the form

$f(x) = a^x$, where $a > 0$ and not equal to 1.

The domain of the function, f, is $(-\infty, +\infty)$. The range is the set of all positive real numbers. If $a < 1$, f is a decreasing function and if $a > 1$ then f is an increasing function. The y-intercept of $f(x)$ is (0, 1) because any base raised to the power of 0 equals 1. Finally, $f(x)$ has a horizontal asymptote at $y = 0$.

<u>Example:</u>   Graph the function $f(x) = 2^x - 4$.

The domain of the function is the set of all real numbers and the range is $y > -4$. Because the base is greater than 1, the function is increasing. The y-intercept of $f(x)$ is (0, -3). The x-intercept of $f(x)$ is (2, 0). The horizontal asymptote of $f(x)$ is $y = -4$.

Finally, to construct the graph of $f(x)$ we find two additional values for the function. For example, $f(-2) = -3.75$ and $f(3) = 4$.

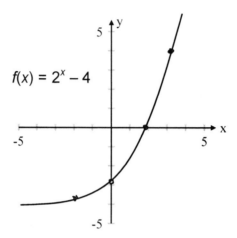

$f(x) = 2^x - 4$

Note that the horizontal asymptote of any exponential function of the form $g(x) = a^x + b$ is $y = b$. Note also that the graph of such exponential functions is the graph of $h(x) = a^x$ shifted b units up or down. Also, the graph of exponential functions of the form $g(x) = a^{(x+b)}$ is the graph of $h(x) = a^x$ shifted b units left or right.

COMPETENCY 8.0   UNDERSTAND THE PROPERTIES AND TECHNIQUES OF ALGEBRAIC OPERATIONS

Skill 8.1   Simplify, evaluate, and perform operations (e.g., factoring, grouping) on polynomials and other algebraic expressions

Simplifying rational expressions

Rational expressions can be simplified into equivalent expressions by reducing. When dividing any number of terms by a single term, divide or reduce their coefficients. Then subtract the exponent of a variable in the denominator from the exponent of the same variable in the numerator.

Example: Simplify

$$\frac{24x^3y^6z^3}{8x^2y^2z} = 3xy^4z^2$$

To reduce a rational expression with more than one term in the denominator, the expression must be factored first. Factors that are exactly the same will cancel and each becomes 1. Factors that have exactly the opposite signs of each other, such as $(a - b)$ and $(b - a)$, will cancel and one factor becomes 1 and the other becomes $a - 1$.

Example: Simplify

$$\frac{3x^2 - 14xy - 5y^2}{x^2 - 25y^2} = \frac{(3x + y)(x - 5y)}{(x + 5y)(x - 5y)} = \frac{3x + y}{x + 5y}$$

Factoring polynomials

To factor a polynomial, follow these steps:

a. Factor out any GCF (greatest common factor)

Example: Factor $36x^4y + 18x^3y^3 - 9x^2y^6$

$$9x^2y(4x^2 + 2xy^2 - y^5)$$

b. For a binomial (2 terms), check to see if the problem is the **difference of perfect squares**. If both factors are perfect squares, then it factors this way:
$$a^2 - b^2 = (a-b)(a+b)$$

If the problem is not the difference of perfect squares, then check to see if the problem is either the sum or difference of perfect cubes. If so, apply one of the following formulae:

$$a^3 + b^3 = (a+b)(a^2 - ab + b^2)$$
$$a^3 - b^3 = (a-b)(a^2 + ab + b^2)$$

Example:

$$x^3 - 8y^3 = (x - 2y)(x^2 + 2xy + 4y^2) \qquad \leftarrow \text{difference}$$

$$64a^3 + 27b^3 = (4a + 3b)(16a^2 - 12ab + 9b^2) \quad \leftarrow \text{sum}$$

c. Trinomials can be perfect squares. Trinomials can be factored into 2 binomials (un-FOILing). Be sure the terms of the trinomial are in descending order. If the last sign of the trinomial is a "+," then the signs in the parentheses will be the same as the sign in front of the second term of the trinomial. If the last sign of the trinomial is a "−," then there will be one "+" and one "−" in the two parentheses. The first term of the trinomial can be factored to equal the first terms of the two factors. The last term of the trinomial can be factored to equal the last terms of the two factors. Work backwards to determine the correct factors to multiply together to get the correct center term.

Example: Factor completely $4x^2 - 25y^2$

No GCF; this is the difference of perfect squares.

$$4x^2 - 25y^2 = (2x - 5y)(2x + 5y)$$

Example: Factor completely $6b^2 - 2b - 8$

GCF of 2; Try to factor into 2 binomials:

$$6b^2 - 2b - 8 = 2(3b^2 - b - 4)$$

Since the last sign of the trinomial is a "−," the signs in the factors are one "+," and one "−." $3b^2$ factors into $3b$ and b. Find factors of 4: 1 & 4; 2 & 2.

$$6b^2 - 2b - 8 = 2(3b^2 - b - 4) = 2(3b - 4)(b + 1)$$

Skill 8.2   **Use number properties (e.g., distributive) to justify algebraic manipulations**

The following properties of real numbers (where a,b,c,d are any real numbers) are useful for manipulation of algebraic expressions.

$a = a$	Reflexive Property
$a + b$ is a unique real number	Closure Property of Addition
ab is a unique real number	Closure Property of Multiplication
If $a = b$, then $b = a$	Symmetric Property
If $a = b$ and $b = c$, then $a = c$	Transitive Property
If $a + b = c$ and $b = d,$ then $a + d = c$	Substitution Property
If $a = b,$ then $a - c = b - c$	Substitution Property
If $a = b,$ then $ac = bc$	Multiplication Property
If $a = b$ and $c \neq 0$, then $\dfrac{a}{c} = \dfrac{b}{c}$	Division Property
$a + b = b + a$	Property of Addition
$ab = ba$	Property of Multiplication
$a + (b + c) = (a + b) + c$	Associative Property of Addition
$a(bc) = (ab)c$	Associative Property of Multiplication
$a + 0 = 0 + a = a$	Identity Property of Addition; the number 0 is called the additive identity
$a(1) = 1(a) = a$	Identity Property of Multiplication; the number 1 is called the multiplicative identity
$a(b + c) = a(b) + a(c)$ and $(b + c)\, a = b(a) + c(a)$	Distributive Property

Example:

$2a + 4 = a - 2$

$2a + 4 - a = a - 2 - a$ Subtract a from both sides

$a + 4 = -2$

$a + 4 - 4 = -2 - 4$ Subtract 4 from both sides

$a = -6$

Check the solution by substitution

$2(-6) + 4 = -6 - 2$

$-12 + 4 = -8$

$-8 = -8$

Example:

$3(2x + 4) - 3(x - 2) = 8x - (7x - 22)$

$3(2x) + 3(4) - 3x - 3(-2) = 8x - 1x - 1(-22)$ Use distributive property three times

$6x = 12 - 3x + 6 = 8x - 7x = 22$

$3x + 18 = x + 22$ Sum up like terms

$3x + 18 - 18 = x + 22 - 18$ Subtract 18 from both sides

$3x = x + 4$

$3x - x = x + 4 - x$ Subtract 18 from both sides

$2x = 4$

$\dfrac{2x}{2} = \dfrac{4}{2}$

$x = 2$

Check by substitution

$3[2(2) + 4] - 3(2 - 2) = 8(2) - [7(2) - 22]$

$3(8) - 3(0) = 16 - (14 - 22)$

$24 - 0 = 16 - (-8)$

$24 = 16 + 8$

$24 = 24$

Skill 8.3 **Solve algebraic equations including problems involving absolute values**

Solving algebraic equations

The procedure for solving algebraic equations is demonstrated using the example below.

Example: $3(x + 3) = {}^-2x + 4$ Solve for x.

1) Expand to eliminate all parentheses.

$$3x + 9 = {}^-2x + 4$$

2) Multiply each term by the LCD to eliminate all denominators (there are none here).

3) Combine like terms on each side when possible (there is no need to do that here).

4) Use real number properties to put all variables on one side and all constants on the other.

$$\rightarrow 3x + 9 - 9 = {}^-2x + 4 - 9 \quad \text{(subtract nine from both sides)}$$

$$\rightarrow 3x = {}^-2x - 5$$

$$\rightarrow 3x + 2x = {}^-2x + 2x - 5 \quad \text{(add 2x to both sides)}$$

$$\rightarrow 5x = {}^-5$$

$$\rightarrow \frac{5x}{5} = \frac{{}^-5}{5} \quad \text{(divide both sides by 5)}$$

$$\rightarrow x = {}^-1$$

Example: Solve $3(2x+5)-4x = 5(x+9)$

$$6x+15-4x = 5x+45$$
$$2x+15 = 5x+45$$
$$^-3x+15 = 45$$
$$^-3x = 30$$
$$x = ^-10$$

Absolute value equations

If a and b are real numbers, and k is a non-negative real number, the solution of $|ax+b| = k$ is $ax+b = k$ and $ax+b = ^-k$

Example: $|2x+3| = 9$ solve for x.

$2x+3 = 9$	and	$2x+3 = ^-9$
$2x+3-3 = 9-3$	and	$2x+3-3 = ^-9-3$
$2x = 6$	and	$2x = ^-12$
$\dfrac{2x}{2} = \dfrac{6}{2}$	and	$\dfrac{2x}{2} = \dfrac{^-12}{2}$
$x = 3$	and	$x = ^-6$

Therefore, the solution is $x = \{3, ^-6\}$

Skill 8.4 **Translate verbal descriptions into algebraic expressions that model problem situations**

Algebraic equations are often used to model and solve real life problems.

Example: Mark and Mike are twins. Three times Mark's age plus four equals four times Mike's age minus 14. How old are the boys?

Since the boys are twins, their ages are the same. "Translate" the English into algebra.

Let x = their age
$3x + 4 = 4x - 14$
$18 = x$

The boys are each 18 years old.

Example: The YMCA wants to sell raffle tickets to raise $32,000. If they must pay $7,250 in expenses and prizes out of the money collected from the tickets, how many tickets worth $25 each must they sell?

Let x = number of tickets sold
Then $25x$ = total money collected for x tickets

Total money minus expenses must be equal to $32,000.
$25x - 7,250 = 32,000$
$25x = 39,250$
$x = 1,570$

If they sell 1,570 tickets, they will raise $32,000.

Example: The Simpsons went out for dinner. All 4 of them ordered the aardvark steak dinner. Bert paid for the 4 meals and included a tip of $12 for a total of $84.60. How much was an aardvark steak dinner?

Let x = the price of one aardvark dinner

So $4x$ = the price of 4 aardvark dinners

$4x = 84.60 - 12$

$4x = 72.60$

$x = \dfrac{72.60}{4} = \$18.15$ The price of one aardvark dinner.

See Skill 9.5 for word problems that use more than one variable.

Skill 8.5 **Model and solve a variety of problems that involve ratio and proportion**

A **ratio** is a comparison of two numbers. If a class had 11 boys and 14 girls, the ratio of boys to girls could be written one of three ways:

11:14   or   11 to 14   or $\dfrac{11}{14}$

The ratio of girls to boys is:

14:11, 14 to 11 or $\dfrac{14}{11}$

Ratios can be reduced when possible. A ratio of 12 cats to 18 dogs would reduce to 2:3, 2 to 3 or $2/3$.

Note: Read ratio questions carefully.

Example: Given a group of 6 adults and 5 children, what is the ratio of children to the entire group?

5:11

A **proportion** is an equation in which a ratio is set equal to another in order to find an unknown quantity in one of the ratios. To solve the proportion, multiply each numerator times the denominator on the other side of the equality sign and set these two products equal to each other. This is called **cross-multiplying** the proportion.

Example: Solve $\dfrac{4}{15} = \dfrac{x}{60}$

To solve this, cross multiply.

$(4)(60) = (15)(x)$

$240 = 15x$

$16 = x$

Example: Solve $\dfrac{x+3}{3x+4} = \dfrac{2}{5}$

To solve, cross multiply.

$5(x+3) = 2(3x+4)$

$5x + 15 = 6x + 8$

$7 = x$

Example: Solve $\dfrac{x+2}{8} = \dfrac{2}{x-4}$

To solve, cross multiply.

$(x+2)(x-4) = 8(2)$

$x^2 - 2x - 8 = 16$

$x^2 - 2x - 24 = 0$

$(x-6)(x+4) = 0$

$x = 6$ or $x = {}^-4$

Proportions can be used to solve word problems whenever relationships are compared. Some situations include scale drawings and maps, similar polygons, speed, time and distance, cost, and comparison shopping.

Example: Which is the better buy, 6 items for $1.29 or 8 items for $1.69?

Find the unit price.

$\dfrac{6}{1.29} = \dfrac{1}{x}$ $\qquad\qquad$ $\dfrac{8}{1.69} = \dfrac{1}{x}$

$6x = 1.29$ $\qquad\qquad\qquad$ $8x = 1.69$

$x = 0.215$ $\qquad\qquad\qquad$ $x = 0.21125$

Thus, 8 items for $1.69 is the better buy.

Example: A car travels 125 miles in 2.5 hours. How far will it travel in 6 hours?

Write a proportion comparing the distance and time.

$$\frac{miles}{hours} \qquad \frac{125}{2.5} = \frac{x}{6}$$
$$2.5x = 750$$
$$x = 300$$

Thus, the car can travel 300 miles in 6 hours.

Example: The scale on a map is $\frac{3}{4}$ inch = 6 miles. What is the actual distance between two cities if they are $1\frac{1}{2}$ inches apart on the map?

Write a proportion comparing the scale to the actual distance.

scale   actual

$$\frac{\frac{3}{4}}{1\frac{1}{2}} = \frac{6}{x}$$

$$\frac{3}{4}x = 1\frac{1}{2}(6)$$

$$\frac{3}{4}x = 9$$

$$x = 12$$

Thus, the actual distance between the cities is 12 miles.

COMPETENCY 9.0 UNDERSTAND LINEAR FUNCTIONS AND THEIR APPLICATIONS

Skill 9.1   Distinguish between linear and nonlinear functions

The individual data points on the graph of a linear relationship cluster around a line of best fit. In other words, a relationship is linear if we can sketch a straight line that roughly fits the data points. Thus, in linear relationships the *y* variable varies by a fixed amount for a fixed change in *x* (e.g. y changes 3 units each time for a unit change in x). Consider the following examples of linear and non-linear relationships.

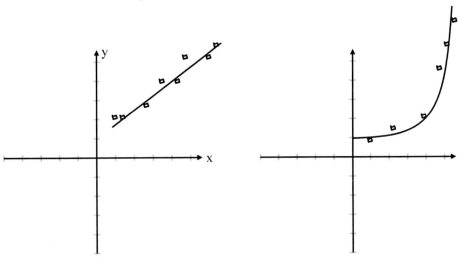

Linear Relationship   Non-Linear Relationship

Note that the non-linear relationship, an exponential relationship in this case, appears linear in parts of the curve.

See Skill 7.5 for more information on non-linear functions.

Skill 9.2 **Analyze the relationship between a linear equation or inequality and its graph**

A relationship between two quantities can be shown using a table, graph, written description or symbolic rule. In the following example, the rule $y = 9x$ describes the relationship between the total amount earned, y, and the number of sunglasses sold, x.

A table using this data would appear as:

number of sunglasses sold	1	5	10	15
total dollars earned	9	45	90	135

Each (x, y) relationship between a pair of values is called the coordinate pair and can be plotted on a graph. The coordinate pairs (1, 9), (5, 45), (10, 90), and (15, 135), are plotted on the graph below.

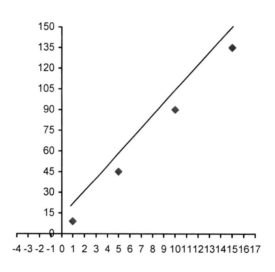

The graph shows a linear relationship. A linear relationship is one in which two quantities are proportional to each other. On a graph, a straight line depicts a linear relationship.

The function or relationship between two quantities may be analyzed to determine how one quantity depends on the other.

For example, the function below shows a linear relationship between y and x: $y = 2x + 1$. The function $y = 2x + 1$, is written as a symbolic rule. The same relationship is also shown in the table below:

x	0	2	3	6	9
y	1	5	7	13	19

The function can also be graphed, as shown:

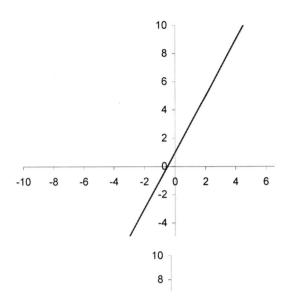

A relationship could be written in words by saying "The value of *y* is equal to two times the value of *x*, plus one." This relationship could be shown on a graph by plotting given points such as the ones shown in the table above.

Another way to describe a function is as a process in which one or more numbers are input into an imaginary machine that produces another number as the output. If 5 is input (*x*) into a machine with a process of *x* + 1, the output (*y*) will equal 6.

In real situations, relationships can be described mathematically. The function, *y* = *x* + 1, can be used to describe the idea that people age one year on their birthday. To describe the relationship in which a person's monthly medical costs are 6 times a person's age, we could write *y* = 6*x*. The monthly cost of medical care could be predicted using this function. A 20 year-old person would spend $120 per month (120 = 20 × 6). An 80 year-old person would spend $480 per month (480 = 80 × 6). Therefore, one could analyze the relationship to say: as you get older, medical costs increase $6.00 per month each year.

Linear inequalities are solved following a procedure similar to that used for solving linear equations. There is however one important point that must be noted while solving inequalities: **dividing or multiplying by a negative number will reverse the direction of the inequality sign.**

The solution to an inequality with one variable is represented in graphical form on the number line or in interval form. In identifying word problems that can be represented by inequalities watch for words like greater than, less than, at least, or no more than.

<u>Example:</u> Solve the inequality. Show its solution using interval form and graph the solution on the number line.

$$\frac{5x}{8} + 3 \geq 2x - 5$$

$$8\left(\frac{5x}{8}\right) + 8(3) \geq 8(2x) - 5(8) \qquad \text{Multiply by LCD = 8.}$$

$$5x + 24 \geq 16x - 40$$

$$5x + 24 - 24 - 16x \geq 16x - 16x - 40 - 24 \quad \text{Subtract 16}x \text{ and 24 from both sides of the equation.}$$

$$^-11x \geq ^-64$$

$$\frac{^-11x}{^-11} \leq \frac{^-64}{^-11}$$

$$x \leq \frac{64}{11} \ ; \ x \leq 5\frac{9}{11}$$

Note the change in direction of the equality with division by a negative number.

Solution in interval form: $\left(^-\infty, 5\frac{9}{11}\right]$

Note: "] " means $5\frac{9}{11}$ is included in the solution.

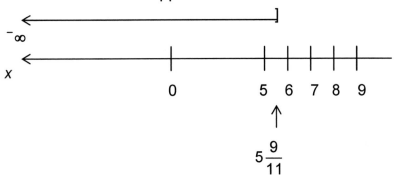

Example: Solve the following inequality and express your answer in both interval and graphical form.

$$3x - 8 < 2(3x - 1)$$

$$3x - 8 < 6x - 2 \qquad\qquad \text{Distributive property.}$$

$$3x - 6x - 8 + 8 < 6x - 6x - 2 + 8$$

Add 8 and subtract $6x$ from both sides of the equation.

$$^{-}3x < 6$$

$$\frac{^{-}3x}{^{-}3} > \frac{6}{^{-}3} \qquad\qquad x >^{-} 2$$

Graphical form:

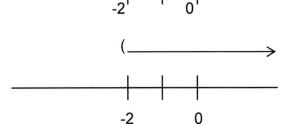

or

Interval form: $(^-2, \infty)$

Recall: a) Using a parentheses or an open circle implies the point in not included in the answer.

b) Using a bracket or a closed circle implies the point is included in the answer.

<u>Example:</u> Solve $6x + 21 < 8x + 31$

$$^-2x + 21 < 31$$

$$^-2x < 10$$

$$x > {}^-5$$

Note that the inequality sign has changed.

To graph an inequality involving two variables x and y, solve the inequality for y. This puts the inequality in the **slope-intercept form**, (for example: $y < mx + b$). The point $(0, b)$ is the y-intercept and m is the line's slope.

When graphing a linear inequality, the line will be dashed if the inequality sign is $<$ or $>$. If the inequality signs are either $\geq$ or $\leq$, the line on the graph will be a solid line. Shade above the line when the inequality sign is $\geq$ or $>$. Shade below the line when the inequality sign is $<$ or $\leq$.

<u>Example:</u> Graph the following inequality.

$$3x - 2y \geq 6$$
$$y \leq 3/2\,x - 3$$

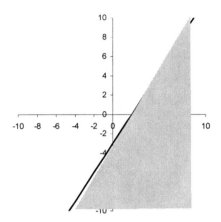

If the inequality solves to x >, $\geq$, < **or** $\leq$ **any number**, then the graph includes a **vertical line**.

If the inequality solves to y >, $\geq$, < **or** $\leq$ **any number**, then the graph includes a **horizontal line**.

For inequalities of the forms $x >$ number, $x \leq$ number , $x <$ number ,or $x \geq$ number, draw a vertical line (solid or dashed). Shade to the right for > or $\geq$. Shade to the left for < or $\leq$.

Example: Graph the following inequality:

$$3x + 12 < -3$$

$$x < {}^-5$$

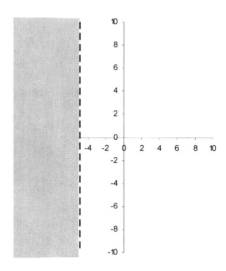

Skill 9.3   **Describe and use various representations (e.g., verbal, tabular, graphical, algebraic) of linear functions**

A linear function can be represented four different ways: verbally, algebraically, graphically, and as a table. Each representation has its advantages and disadvantages. For example, the graph of a line is the best visual representation of a linear relationship: it displays the slope, *x*- and *y*-intercepts (if any), and other characteristics. However, the graph of a line is not always the best representation to find values for *x* and *y* not on shown on the graph.

For example, for every pound an object weighs on Earth, the object weighs 1.4 pounds on Neptune. This is a verbal representation of a linear function. The function can be written algebraically: $y = 1.4x$. You can also use the algebraic function to make a table of values:

x	y
1	1.4
2	2.8
3	4.2
4	5.6
5	7.0

Finally, you can use the values from the table to draw the graph of the line.

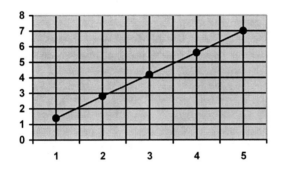

Example: Mr. Smith is starting a business selling gift baskets. He spent $700 for supplies and materials. He earns a profit of $21 for each basket he sells. Write this relationship algebraically, as a table of values, and as a graph.

The relationship is a linear function that can be written algebraically: $y = 21x - 700$

A table of values is shown below:

x	y
0	-700
10	-490
20	-280
30	-70
40	140

The graph of the line is shown below:

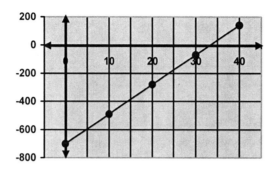

See Skill 9.2 for more examples.

Skill 9.4   **Select a linear equation that best fits a set of data**

A linear function is a function whose graph is a straight line. It has the algebraic form $y = ax + b$. The variables a and b are constants. In order to determine the linear function that best models a set of data (x,y), we need to find *a* and *b*.

Example: Johnny delivers flyers for ABC Hardware after school. He is paid a weekly salary of $25 plus a 1% commission on any sales resulting from the flyers. The table of data would look like this:

sales, s	earnings, E(s)
100	26
200	27
300	28
400	29
500	30

The linear function would be $E(s) = (0.01)s + 25$, where s is the sales amount in dollars and *E(s)* represents total earnings. Here *a* = 0.01 and *b = 25*.

In the above example, the data fits the straight line exactly. The term "best fit" is used more commonly in describing a set of data where all the points do not fall exactly on a straight line but show a general linear trend. The best fit straight line in that case is the one that minimizes the standard deviation of the data points from the line and is determined by the method of linear regression (see Skill 10.6).

Example: Tiffany recorded the amount of rain that accumulated in Columbus, GA, over six weeks. Her results are shown below. Write an equation that represents the amount of rain that fell.

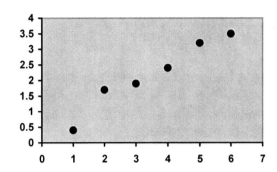

To draw a trend line, or line of best fit, draw a line that comes closest to connecting the points.

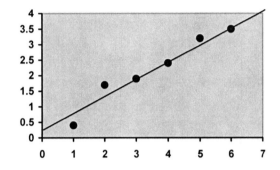

Choose two points on the line to find the slope.

(3, 1.9), (6, 3.5)

$$m = \frac{y_2 - y_1}{x_2 - x_1} = \frac{3.5 - 1.9}{6 - 3} = \frac{1.6}{3} = 0.53$$

Substitute a point into the equation and solve to find the y-intercept.
$y = 0.53x + b$
$1.9 = 0.53(3) + b$
$b = 0.31$
The equation of the line of best fit is $y = 0.53x + 0.31$.

Skill 9.5   **Represent and solve problems involving linear equations and inequalities and systems of linear equations and inequalities algebraically and graphically and interpreting solutions to these problems**

Skills 8.4 and 8.5 provide many examples of problems involving linear equations with one variable.

Problems with more than one variable may be modeled using a system of linear equations, i.e. a group of equations consisting of as many equations as variables. Examples of systems of equations with two variables are given below.

The solution set of a **pair of linear equations** is all the ordered pairs of real numbers that satisfy both equations. Graphically, the solution represents the intersection of the lines.

Equivalent or **dependent** equations are two equations that are represented by the same line; that is, one is a multiple of the other.

Example: $3x - 4y = 8$
                         $6x - 8y = 16$

Consistent equations are pairs of equations to which we can find a solution.

Example: $x + 3y = 5$
                         $2x - y = 15$

Inconsistent equations can be represented by parallel lines; that is, there are no values of x and y that satisfy both equations.

Example: $2x - y = 3$
                         $6x - 3y = 2$

There are several methods for solving linear equations. Three methods include **graphing**, **linear combinations**, and **substitution**.

In the **substitution** method, an equation is solved for either variable. Then, that solution is substituted in the other equation to find the remaining variable.

Example: (1) $2x + 8y = 4$
          (2) $x - 3y = 5$

          (2a) $x = 3y + 5$ Solve equation (2) for x

          (1a) $2(3y + 5) + 8y = 4$ Substitute x in equation (1)
                 $6y + 10 + 8y = 4$ Solve.
                 $14y = -6$
                 $y = \frac{-3}{7}$ Solution

          (2) $x - 3y = 5$
                $x - 3(\frac{-3}{7}) = 5$ Substitute the value of y.
                $x = \frac{26}{7} = 3\frac{5}{7}$ Solution

Thus the solution set of the system of equations is $(3\frac{5}{7}, \frac{-3}{7})$.

In the **linear combinations** or **addition-subtraction** method, one or both of the equations are replaced with an equivalent equation in order that the two equations can be combined (added or subtracted) to eliminate one variable.

Example: (1) $4x + 3y = -2$
          (2) $5x - y = 7$

          (1) $4x + 3y = -2$
          (2a) $15x - 3y = 21$ Multiply equation (2) by 3

                $19x = 19$ Combining (1) and (2a)
                 $x = 1$ Solve.

To find y, substitute the value of x in equation (1) or (2).
          (1) $4x + 3y = -2$
                $4(1) + 3y = -2$
                $4 + 3y = -2$
                $3y = -2 - 4$
                $y = -2$

Thus the solution is $x = 1$ and $y = -2$ or the ordered pair is (1, -2).

Example: Solve for x and y.

$$4x + 6y = 340$$
$$3x + 8y = 360$$

Multiply the first equation by 4: $4(4x + 6y = 340)$

Multiply the second equation by $^-3$: $^-3(3x + 8y = 360)$

The equations can now be added to each other to eliminate one variable and solve for the other variable.

$$16x + 24y = 1360$$
$$-9x - 24y = {}^-1080$$
$$7x = 280$$
$$x = 40$$

Substituting the value of x in the first equation and solving for y,
$$4(40) + 6y = 340$$
$$6y = 180$$
$$y = 30$$

Some word problems can be solved using a system of equations.

Example: Farmer Greenjeans bought 4 cows and 6 sheep for $1,700. Mr. Ziffel bought 3 cows and 12 sheep for $2,400. If all the cows were the same price and all the sheep were another price, find the price charged for a cow or for a sheep.

Let x = price of a cow
Let y = price of a sheep

Then Farmer Greenjeans' equation would be: $4x + 6y = 1700$
Mr. Ziffel's equation would be: $3x + 12y = 2400$

To solve by **addition-subtraction**:

Multiply the first equation by $^-2$: $^-2(4x+6y=1700)$
Keep the other equation the same : $(3x+12y=2400)$

By doing this, the equations can be added to each other to eliminate one variable and solve for the other variable.

$$^-8x-12y = ^-3400$$
$$\underline{3x+12y = 2400}$$ Add these equations.
$$^-5x \quad\quad = ^-1000$$

$x = 200 \leftarrow$ the price of a cow was \$200.
Solving for y, $y=150 \leftarrow$ the price of a sheep,\$150.

To solve by **substitution**:

Solve one of the equations for a variable. (Try to make an equation without fractions if possible.) Substitute this expression into the equation that you have not yet used. Solve the resulting equation for the value of the remaining variable.

$$4x+6y=1700$$
$$3x+12y=2400 \leftarrow \text{ Solve this equation for } x.$$

It becomes $x = 800-4y$. Now substitute $800-4y$ in place of x in the other equation. $4x+6y=1700$ now becomes:

$$4(800-4y)+6y=1700$$
$$3200-16y+6y=1700$$
$$3200-10y=1700$$
$$^-10y = ^-1500$$
$$y=150, \text{ or \$150 for a sheep.}$$

Substituting 150 back into an equation for y, find x.
$$4x+6(150)=1700$$
$$4x+900=1700$$
$$4x=800 \text{ so } x=200 \text{ for a cow.}$$

To solve a system of equations by **graphing**, draw the graphs of the lines. The solution is the point where the lines intersect.

Example: What is the solution to the system of equations below?

$$y = 3x - 4$$
$$y = -x + 7$$

Make a table of values for each equation and graph the lines on the same graph.

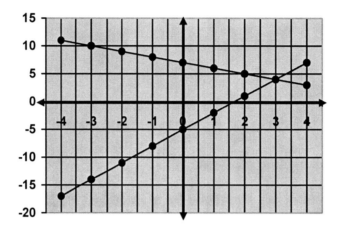

The solution is (3, 4), the point where both lines intersect.

The solution to a system of linear inequalities consists of the part of the graph where the shaded areas for all the inequalities in the system overlap. For instance, if the graph of one inequality was shaded with red, and the graph of another inequality was shaded with blue, then the overlapping area would be shaded purple. The points in the purple area would be the solution set of this system.

<u>Example:</u> Solve by graphing:

$$x + y \leq 6$$
$$x - 2y \leq 6$$

Solving the inequalities for y, they become:

$$y \leq {}^-x + 6 \quad (y\text{-intercept of 6 and slope} = {}^-1)$$

$$y \geq 1/2\, x - 3 \quad (y\text{-intercept of } {}^-3 \text{ and slope} = 1/2\,)$$

A graph with shading is shown below:

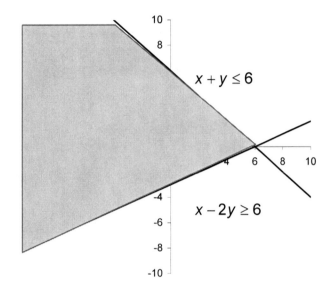

Example: Sharon's Bike Shoppe can assemble a 3-speed bike in 30 minutes or a 10-speed bike in 60 minutes. The profit per bike sold is $60 for a 3-speed and $75 for a 10-speed bike. Total daily profit must be at least $300. How many of each type of bike should they assemble during an 8 hour (480 minute) day to make the maximum profit?

> Let x = number of 3-speed bikes.
> y = number of 10-speed bikes.

Since there are only 480 minutes to use each day,

$30x + 60y \leq 480$ is the first inequality.

Since the total daily profit must be at least $300,

$60x + 75y \geq 300$ is the second inequality.

Solving for y:
> $30x + 60y \leq 480$ becomes $y \leq 8 - 1/2\,x$
> $60y \leq -30x + 480$
>
> $$y \leq -\frac{1}{2}x + 8$$

Solving for y:
> $60x + 75y \geq 300$ becomes $y \geq 4 - 4/5\,x$
> $75y + 60x \geq 300$
>
> $$75y \geq -60x + 300$$
>
> $$y \geq -\frac{4}{5}x + 4$$

Graph these 2 inequalities:
$$y \leq 8 - 1/2\,x$$
$$y \geq 4 - 4/5\,x$$

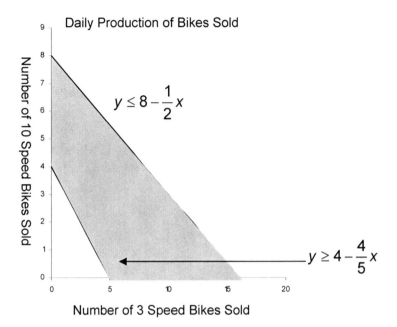

Daily Production of Bikes Sold

$$y \leq 8 - \frac{1}{2}x$$

$$y \geq 4 - \frac{4}{5}x$$

Number of 10 Speed Bikes Sold

Number of 3 Speed Bikes Sold

Notice $x \geq 0$ and $y \geq 0$, because the number of bikes assembled can not be a negative number. Graph these as additional constraints on the problem. The number of bikes assembled must always be an integer value, so points within the shaded area of the graph must have integer values. The maximum profit will occur at or near a corner of the shaded portion of this graph. Those points occur at (0,4), (0,8), (16,0), or (5,0).

Using the coordinates of the graph's corners and the profits of $60/3$-speed or $75/10$-speed, the daily profits could be:

$(0,4)$ $\$60(0) + \$75(4) = \$300$

$(0,8)$ $\$60(0) + \$75(8) = \$600$

$(16,0)$ $\$60(16) + \$75(0) = \$960 \leftarrow$ Maximum profit

$(5,0)$ $\$60(5) + \$75(0) = \$300$

The maximum profit would occur if 16 3-speed bikes and no 10-speed bikes are made daily.

Skill 9.6   **Interpret the meaning of the slope and y-intercept in a given situation**

To find the slope of a line, solve a linear equation for y to get it into the **slope-intercept form**, $y = mx + b$ where m is the slope and b is the y-intercept, i.e. the y-coordinate of the point where the line crosses the y-axis. To find the y-intercept, substitute 0 for x and solve for y. The x-intercept is the x-coordinate of the point where the line crosses the x-axis. To find the x-intercept, substitute 0 for y and solve for x.

The graph below represents the following equation (notice that the y-intercept = 3 and the x-intercept=6/5):

$$5x + 2y = 6$$
$$y = {}^-5/2\,x + 3$$

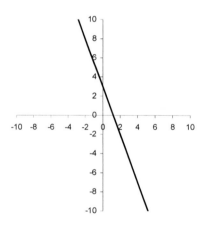

If an equation is of the form **x = any number**, its graph is a **vertical line** and it only has an x-intercept. Its slope is **undefined**.

If an equation is of the form **y = any number**, its graph is a **horizontal line** and it only has a y-intercept. Its slope is 0.

The equation of a straight line can be found from its graph by finding the y-intercept and slope (see Skill 6.2 for the slope formula).

Example: Find the equation of a line that passes through the points $(9, {}^-6)$ and $({}^-1, 2)$.

$$\text{slope} = \frac{y_2 - y_1}{x_2 - x_1} = \frac{2 - {}^-6}{{}^-1 - 9} = \frac{8}{{}^-10} = \frac{{}^-4}{5}$$

The y-intercept may be found by substituting the slope (*m*) and the coordinates (x, y) for one of the data points in the slope-intercept form of the equation $y = mx + b$ giving

$$-6 = -\frac{4}{5} \times 9 + b \quad \text{where b is the y-intercept}$$

$$b = \frac{6}{5}$$

Thus the slope-intercept form of the equation is

$$y = -\frac{4}{5}x + \frac{6}{5}$$

Multiplying by 5 to eliminate fractions, it is:

$$5y = {}^-4x + 6 \rightarrow 4x + 5y = 6 \quad \text{Standard form.}$$

Example: Find the slope and intercepts of the linear equation $3x + 2y = 14$.

$$3x + 2y = 14$$
$$2y = {}^-3x + 14$$
$$y = {}^-3/2\, x + 7$$

The slope of the line is ${}^-3/2$. The *y*-intercept of the line is 7.

The intercepts can also be found by substituting 0 in place of the other variable in the equation.

To find the y-intercept:
let $x = 0$; $3(0) + 2y = 14$
$0 + 2y = 14$
$2y = 14$
$y = 7$
$(0,7)$ is the y-intercept.

To find the x-intercept:
let $y = 0$; $3x + 2(0) = 14$
$3x + 0 = 14$
$3x = 14$
$x = 14/3$
$(14/3, 0)$ is the x intercept.

Example: Sketch the graph of the line represented by $2x + 3y = 6$.

Let $x = 0 \rightarrow 2(0) + 3y = 6$
$\rightarrow 3y = 6$
$\rightarrow y = 2$
$\rightarrow (0,2)$ is the y-intercept.

Let $y = 0 \rightarrow 2x + 3(0) = 6$
$\rightarrow 2x = 6$
$\rightarrow x = 3$
$\rightarrow (3,0)$ is the x-intercept.

Let $x = 1 \rightarrow 2(1) + 3y = 6$
$\rightarrow 2 + 3y = 6$
$\rightarrow 3y = 4$
$\rightarrow y = \dfrac{4}{3}$
$\rightarrow \left(1, \dfrac{4}{3}\right)$ is the third point.

Plotting the three points on the coordinate grid, we get the following:

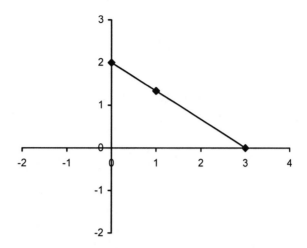

Example: Juan recorded the weight of his puppy for 4 weeks. He discovered that the weight of the puppy, in pounds, can be represented by the function $y = 2.2x + 4.5$. What do the slope and y-intercept of the equation represent?

The slope represents the number of pounds the puppy is gaining each week. In this situation, the puppy is gaining 2.2 pounds per week. The y-intercept represents the weight of the puppy before the first week Juan began recording the puppy's weight. Thus, the puppy weighed 4.5 before the first week.

SUBAREA IV.   **DATA ANALYSIS AND PROBABILITY**

COMPETENCY 10.0 UNDERSTAND THE PROCESS OF COLLECTING, ORGANIZING, DESCRIBING AND INTERPRETING DATA

Skill 10.1   **Demonstrate knowledge of the nature of sampling, the collection of data through surveys and experiments, the importance of sample size, and random sampling**

In cases where the number of events or individuals is too large to collect data on each one, scientists collect information from only a small percentage. This is known as **sampling** or **surveying**. If sampling is done correctly, it should give the investigator nearly the same information he would have obtained by testing the entire population. The survey must be carefully designed, considering both the sampling technique and the size of the sample.

Bias occurs in a sample when some members or opinions of a population are less likely to be included than others. The method a survey is taken can contribute to bias in a survey.

There are a variety of sampling techniques: random, systematic, stratified, cluster, and quota are just a few. A truly **random** sample must choose events or individuals without regard to time, place, or result. Random samples are least likely to be biased because they are most likely to represent the population from which they are taken. **Stratified, quota,** and **cluster** sampling all involve the definition of sub-populations. Those subpopulations are then sampled randomly in an attempt to represent many segments of a data population evenly. While random sampling is typically viewed as the "gold standard", sometimes compromises must be made to save time, money, or effort. For instance, when conducting a phone survey, calls are typically only made in a certain geographical area and at a certain time of day. This is an example of cluster sampling. There are three stages to cluster or area sampling: the target population is divided into many regional clusters (groups); a few clusters are randomly selected for study; a few subjects are randomly chosen from within a cluster

Systematic sampling involves the collection of a sample at defined intervals (for instance, every 10th part to come off a manufacturing line).

Convenience sampling is the method of choosing items arbitrarily and in an unstructured manner from the frame. Convenience samples are most likely to be biased because they are likely to exclude some members of a population.

Another important consideration in sampling is sample size. Again, a large sample will yield the most accurate information but other factors often limit sample size. Statistical methods may be used to determine how large a sample is necessary to give an investigator a specified level of certainty (95% is a typical confidence interval).

Conversely, if a scientist has a sample of certain size, those same statistical methods can be used to determine how confident the scientist can be that the sample accurately reflects the whole population. The smaller the sample size, the more likely the sample is biased.

Example: Brittany called 500 different phone numbers from the phone book to ask people which candidate they were voting for. Which type of sample did Brittany use? Is the sample biased?

Brittany used a random sample. The sample is not biased because it is random and the sample size is appropriate.

Example: Jacob surveyed the girls' softball team on their favorite foods. Which type of sample did he use? Is the sample biased?

Jacob used a convenience sample. The sample is biased because it only sampled a small population of girls.

Skill 10.2 Select an appropriate format for organizing and displaying data

Often data is made more readable and user-friendly by consolidating the information in the form of a graph.

Bar graphs are used to compare various quantities using bars of different lengths. A **pictograph** shows comparison of quantities using symbols. Each symbol represents a number of items. To make a **bar graph** or a **pictograph**, determine the scale to be used for the graph. Then determine the length of each bar on the graph or determine the number of pictures needed to represent each item of information. Be sure to include an explanation of the scale in the legend.

Example: A class had the following grades:
4 A's, 9 B's, 8 C's, 1 D, 3 F's.
Graph these on a bar graph and a pictograph.

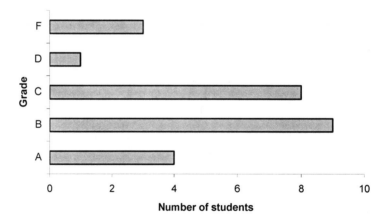

Pictograph

Grade	Number of Students
A	😊😊😊😊
B	😊😊😊😊😊😊😊😊😊
C	😊😊😊😊😊😊😊😊
D	😊
F	😊😊😊

Bar graph

Line graphs are used to show trends, often over a period of time.

To make a line graph, determine appropriate scales for both the vertical and horizontal axes (based on the information to be graphed). Describe what each axis represents and mark the scale periodically on each axis. Graph the individual points of the graph and connect the points on the graph from left to right.

Example: Graph the following information using a line graph.

Height of Two Pea Plants for Six Days

Day	1	2	3	4	5	6
Plant 1 Height (in.)	1.2	1.4	1.4	1.8	2.3	2.4
Plant 2 Height (in.)	0.7	0.9	1.2	1.5	1.6	1.7

Height of Two Pea Plants for Six Days

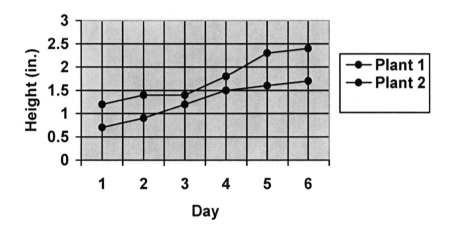

Circle graphs show the relationship of various parts of a data set to each other and the whole. Each part is shown as a percentage of the total and occupies a proportional sector of the circular area. To make a circle graph, total all the information that is to be included on the graph. Determine the central angle to be used for each sector of the graph using the following formula:

$$\frac{\text{information}}{\text{total information}} \times 360° = \text{degrees in central} \sphericalangle$$

Lay out the central angles according to these sizes, label each section and include its percentage.

<u>Example:</u> Graph this information on a circle graph:

Monthly expenses:

Rent, $400
Food, $150
Utilities, $75
Clothes, $75
Church, $100
Misc., $200

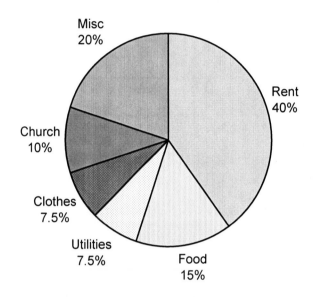

Scatter plots compare two characteristics of the same group of things or people and usually consist of a large body of data. They show how much one variable is affected by another. The relationship between the two variables is their **correlation**. The closer the data points come to making a straight line when plotted, the closer the correlation.

Stem-and-leaf plots are visually similar to histograms. The **stems** are the digits in the greatest place value of the data values, and the **leaves** are the digits in the next greatest place values. Stem and leaf plots are best suited for small sets of data and are especially useful for comparing two sets of data.

Example: Make a stem-and-leaf plot to display the following test scores: 49, 54, 59, 61, 62, 63, 64, 66, 67, 68, 68, 70, 73, 74, 76, 76, 76, 77, 77, 77, 77, 78, 78, 78, 78, 83, 85, 85, 87, 88, 90, 90, 93, 94, 95, 100, 100.

stem	leaves
4	9
5	4 9
6	1 2 3 4 6 7 8 8
7	0 3 4 6 6 6 7 7 7 7 8 8 8 8
8	3 5 5 7 8
9	0 0 3 4 5
10	0 0

Histograms are used to summarize information from large sets of data that can be naturally grouped into intervals. The vertical axis indicates **frequency** (the number of times any particular data value occurs), and the horizontal axis indicates data values or ranges of data values. The number of data values in any interval is the **frequency of the interval**.

Example: The human resources department of a small company surveyed workers on their weekly salaries. Ten workers earned between $600 and $624 per week, Twenty workers earned between $625–$649/week. Twenty workers earned between $650–$674/week. Fifteen workers earned between $675–$699/week. Twenty-five workers earned between $700–$724/week. Five workers earned $725 or more each week. Draw a histogram to display the results.

Employees' Salaries per Week

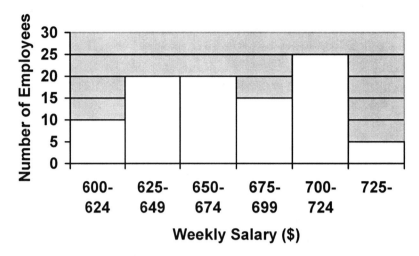

Example: Justin surveyed his classmates on their favorite magazines. What is the best graph to show his data?

Since Justin is using counting data, so a bar graph or pictograph would be the best format to display his data.

Example: Lakeisha recorded the amount of rain that fell in Augusta, GA, for 6 weeks. What is the best graph to show her data?

Since Lakeisha is displaying a change over time, a line graph would be the best format to show her data.

Skill 10.3 **Analyze data in a variety of formats (i.e., tables, frequency distributions, histograms, circle graphs, bar graphs, line plots, scatter plots, stem-and-leaf plots, and box-and whisker plots)**

Displaying data in graphical format can reveal a lot of information about the data set. An **inference** is a statement that is derived from reasoning. When reading a graph, inferences help with interpretation of the data that is being presented. From this information, a **conclusion** and even **predictions** about what the data actually means is possible.

A **trend** line on a line graph shows the correlation between two sets of data. A trend may show positive correlation (both sets of data get greater together) negative correlation (one set of data increases while the other decreases), or no correlation.

<u>Example:</u> Katherine and Tom were both doing poorly in math class. Their teacher had a conference with each of them in November. The following graph shows their math test scores during the school year.

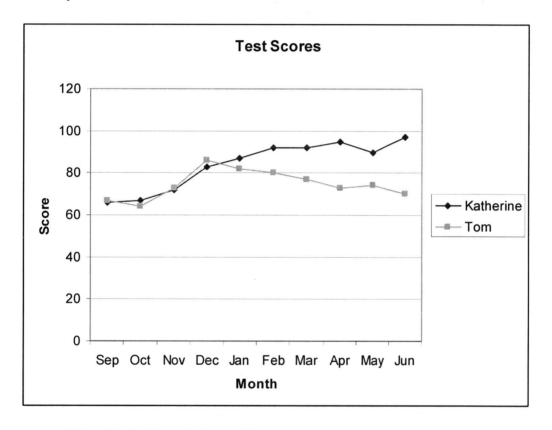

What kind of trend does this graph show?

This graph shows that there is a positive trend in Katherine's test scores and a negative trend in Tom's test scores.

What inferences can you make from this graph?

We can infer that Katherine's test scores rose steadily after November. Tom's test scores spiked in December but then began a negative trend.

What conclusion can you draw based upon this graph?

We can conclude that Katherine took her teacher's meeting seriously and began to study in order to do better on the exams. It seems as though Tom tried harder for a bit but his test scores eventually slipped back down to the level where he began.

See also Skill 10.2. For an example on analyzing a box-and-whisker plot, see Skill 10.5.

Skill 10.4 **Determine and analyze measures of central tendency (i.e., mean and median), mode, and dispersion (e.g., range, variance, standard deviation, interquartile range, outliers)**

The arithmetic **mean** (or average) of a set of numbers is the sum of the numbers given, divided by the number of items being averaged.

Example: Find the mean of the following set of numbers. Round to the nearest tenth.
> 24.6, 57.3, 44.1, 39.8, 64.5
> The mean is 230.3 /5
> = 46.06, rounded to 46.1

The **median** is the middle number in an ordered set of numbers. To calculate the median, the terms must first be arranged in order. If there are an even number of terms, the median is the mean of the two middle terms.

Example: Find the median of the following set of numbers.
> 12,   14,   27,   3,   13,   7,   17,   12, 22,   6,   16

> Rearrange the terms.
> 3,   6,   7,   12,   12,   13,   14,   16,   17,   22,   27

Since there are 11 numbers, the median would be the middle or sixth number, 13.

The **mode** of a set of numbers is the number that occurs with the greatest frequency. A set can have no mode if each term appears exactly one time. Similarly, there can also be more than one mode.

Example: Find the mode of the following set of numbers.

> 26, 15, 37, 26, 35, 26, 15

15 appears twice, but 26 appears 3 times, therefore the mode is 26.

The **range** of a data set is the difference between the highest and lowest value of data items.

The **variance** of a data set is the sum of the squares of the deviation of each data item x from the mean $\overline{x}$ divided by the number of items N. (the lower case Greek letter sigma squared (σ^2) represents variance).

$$\frac{\sum(x-\overline{x})^2}{N} = \sigma^2$$

The larger the value of the variance the larger the spread.

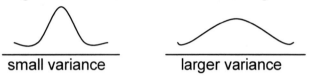

small variance   larger variance

Standard deviation is defined as the square root of the variance. The lower case Greek letter sigma (σ) is used to represent standard deviation.

$$\sigma = \sqrt{\sigma^2}$$

Most statistical calculators have standard deviation keys on them and should be used when asked to calculate statistical functions. It is important to become familiar with the calculator and the location of the keys needed.

<u>Example:</u> Given the ungrouped data below, calculate the mean, range, standard deviation and the variance.

15	22	28	25	34	38
18	25	30	33	19	23

Mean ($\overline{X}$) = 25.8333333
Range: $38 - 15 = 23$
standard deviation (σ) = 6.99137
Variance (σ^2) = 48.87879

Percentiles divide a data set into 100 equal parts. A person whose score falls in the 65th percentile has outperformed 65 percent of all those who took the test. This does not mean that the score was 65 percent out of 100 nor does it mean that 65 percent of the questions answered were correct. It means that the grade was higher than 65 percent of all those who took the test.

<u>Example:</u> Given the following set of data, find the percentile of the score 104.

70, 72, 82, 83, 84, 87, 100, 104, 108, 109, 110, 115

Find the percentage of scores below 104.

7/12 of the scores are less than 104. This is 58.333%; therefore, the score of 104 is in the 58th percentile.

Stanine "standard nine" scores combine the understandability of percentages with the properties of the normal curve of probability. Stanines divide the bell curve into nine sections, the largest of which stretches from the 40th to the 60th percentile and is the "Fifth Stanine" (the average of taking into account error possibilities).

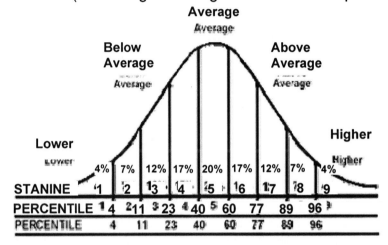

Quartiles divide the data into 4 parts. First find the median of the data set (Q2), then find the median of the upper (Q3) and lower (Q1) halves of the data set. The upper and lower quartiles may be determined in two different ways and statisticians don't agree on which method to use. Tukey's method for finding the quartile values is to find the median of the data set, then find the median of the upper and lower halves of the data set. If there are an odd number of values in the data set, include the median value in both halves when finding the quartile values. For example, if we have the data set:

{1, 4, 9, 16, 25, 36, 49, 64, 81}

First find the median value, which is 25. Since there are an odd number of values in the data set (9), we include the median in both halves. To find the quartile values, we must find the medians of:

{1, 4, 9, 16, 25} and {25, 36, 49, 64, 81}

Since each of these subsets has an odd number of elements (5), we use the middle value. Thus the lower quartile value is 9 and the upper quartile value is 49.

If the test you are taking allows the use of the TI-83 calculator, know that it uses a method described by Moore and McCabe (sometimes referred to as "M-and-M") to find quartile values. Their method is similar to Tukey's, but you *don't* include the median in either half when finding the quartile values. Using M-and-M on the data set above:

$$\{1, 4, 9, 16, 25, 36, 49, 64, 81\}$$

First find that the median value is 25. This time we'll exclude the median from each half. To find the quartile values, we must find the medians of:

$$\{1, 4, 9, 16\} \text{ and } \{36, 49, 64, 81\}$$

Since each of these data sets has an even number of elements (4), we average the middle two values. Thus the lower quartile value is (4+9)/2 = 6.5 and the upper quartile value is (49+64)/2 = 56.5.

With each of the above methods, the quartile values are always either one of the data points, or exactly half way between two data points.

Example: Find the first, second and third quartile for the data listed.
6, 7, 8, 9, 10, 12, 13, 14, 15, 16, 18, 23, 24, 25, 27, 29, 30, 33, 34, 37

Quartile 1: The 1st Quartile is the median of the lower half of the data set, which is 11.

Quartile 2: The median of the data set is the 2nd Quartile, which is 17.

Quartile 3: The 3rd Quartile is the median of the upper half of the data set, which is 28.

An **outlier** is a number in a set of data that is much larger or smaller than most of the other numbers in the set.
Example: In a golf tournament, the following scores were recorded for 10 players: 73, 77, 84, 75, 79, 106, 74, 80, 83, and 79. What is the outlier?

The 106 score is the outlier because is greater than the other scores.

Skill 10.5 **Interpret summary statistics to determine trends and draw valid conclusions from data (e.g., census)**

When trying to interpret large volumes of data, it is frequently difficult to draw conclusions from the raw unprocessed numbers. Therefore, in many situations you will be provided with summary statistics that summarize the study's findings and enable people to draw useful inferences. Common summary statistics include mean, median, mode, standard deviation, variance, and range. These statistics describe various characteristics of the data. The first group, mean, median, and mode, describe the overall magnitude of the distribution. The second group, standard deviation, variance, and range, describe whether the data is tightly clumped together in value or if it is widely dispersed. Together, these two groups of statistics can be used to develop and present conclusions about data from a variety of sources.

When looking at trend data it is important to look carefully at what is being graphed. Read the axes carefully in order to determine the scales involved, and whether the data presented is in the form of raw data or summary statistics. A graph of the average age of the population over time tells you something completely different than the graph of the number of 18 year-olds over time. Understanding what precisely is being shown will greatly enhance your ability to draw conclusions from a graph. The definitions of the quantities used in summary statistics are described in detail in Skill 10.4. Here we discuss the implications and uses of these quantities.

Mean: The numerical average of the data.

Median: The middle value in a data set. This can be a more useful value than the mean if there are large numbers of outliers on one end of the data set. Such values, which fall significantly outside the range of the majority of the data, can bias the calculation of the mean. In a normally distributed data set, the median and mean are effectively identical. If the mean and median are distinctly different, it suggests that the data set is skewed in one direction or another.

Mode: The value that is present the most times in a data set. For example, if you were measuring the heights of students in a classroom and there were 5 students who were 5'6" tall, and the other students were 5'11", 5'8", 5'4", 5'5", and 5'2", the mode would be 5'6".

Variance: The variance looks at, on average, how far each value in the data set is from the mean.

Standard Deviation: The standard deviation defines how wide the spread of the data is. If all the data is tightly clustered around the mean, the standard deviation will be relatively small. If the data has a lot of variation, and isn't tightly clustered, the standard deviation will be large. It is the square root of the variance.

Range: The range is just what it sounds like. It is the difference between the largest and smallest value in the dataset (max − min). It tells you the scope of the data. The interquartile range, in contrast, is the difference between the 25th and 75th percentile of the data set. These are the values that define the box in a box and whisker plot.

A **box and whisker plot** (on the next page) is a common way of visualizing all these summary statistics. The central line of the box is the median value. The top and bottom lines of the box define the 75th and 25th percentile values of the data. The whiskers define the range of the data, although sometimes they may exclude outlying values which would instead be represented by individual points.

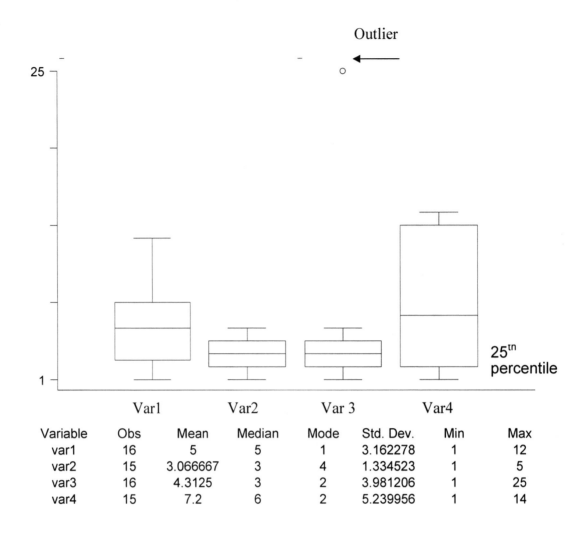

Variable	Obs	Mean	Median	Mode	Std. Dev.	Min	Max
var1	16	5	5	1	3.162278	1	12
var2	15	3.066667	3	4	1.334523	1	5
var3	16	4.3125	3	2	3.981206	1	25
var4	15	7.2	6	2	5.239956	1	14

Skill 10.6 **Analyze and draw conclusions about the relationship between two variables**

Correlation is a measure of the association between two variables. The **correlation coefficient** (r) is used to describe the strength of the association between the variables and the direction of the association. It varies from -1 to 1, with 0 being a random relationship, 1 being a perfect positive linear relationship, and -1 being a perfect negative linear relationship.

Example:

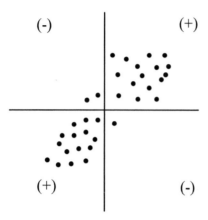

Horizontal and vertical lines are drawn through the point of averages of the x and y values. This divides the scatter plot into four quadrants. If a point is in the lower left quadrant, the product of two negatives is positive; in the upper right, the product of two positives is positive. The positive quadrants are depicted with the positive sign (+). In the two remaining quadrants (upper left and lower right), the product of a negative and a positive is negative. The negative quadrants are depicted with the negative sign (-). If r is positive, then there are more points in the positive quadrants and if r is negative, then there are more points in the two negative quadrants.

Regression is a form of statistical analysis used to predict a dependent variable (y) from values of an independent variable (x). There are many different techniques for performing regression analysis and they will not be discussed here. It is enough to say that the end product of regression is a regression equation that is derived from a known set of data and is seen as the best way to express in equation form the x-y relationship demonstrated by the data.

The simplest regression analysis models the relationship between two variables using the following equation: $y = a + bx$, where y is the dependent variable and x is the independent variable. This simple equation denotes a linear relationship between x and y. This form would be appropriate if, when you plotted a graph of x and y, you tended to see the points roughly form along a straight line. The line can then be used to make predictions of new data values.

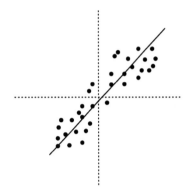

If all of the data points fell on the line, there would be a perfect correlation ($r = 1.0$) between the x and y data points. These cases represent the best scenarios for prediction. A positive or negative r value represents how y varies with x. When r is positive, y increases as x increases. When r is negative y decreases as x increases.

A **linear regression** equation is of the form: $Y = a + bX$.

<u>Example:</u> A teacher wanted to determine how a practice test influenced a student's performance on the actual test. The practice test grade and the subsequent actual test grade for each student are given in the table below:

Practice Test (x)	Actual Test (y)
94	98
95	94
92	95
87	89
82	85
80	78
75	73
65	67
50	45
20	40

We determine the equation for the linear regression line to be
$$y = 14.650 + 0.834x.$$

A new student comes into the class and scores 78 on the practice test. Based on the equation obtained above, what would the teacher predict this student would get on the actual test?

$$y = 14.650 + 0.834(78)$$

$$y = 14.650 + 65.052$$

$$y = 80 = 79.702$$

COMPETENCY 11.0 UNDERSTAND THE THEORY AND APPLICATIONS OF PROBABILITY

Skill 11.1   Apply basic concepts of set theory (e.g., subsets, complements, unions)

A set is a collection of objects or elements that are not ordered. Elements of a set are not repeated.

Examples: {a, b, c, d, e} or {10, 20, 30, 40....100} or {All animals found in Africa}

A **subset** of a set consists of some of the elements in the set. If B is a subset of A, then every element of B is included in A. This is written as $B \subseteq A$. A **proper subset** of a set consists of some of the elements in a set, never all of them. If C is a proper subset of A then every element of C is included in A and A has at least one element that is not in C. The proper subset relationship is written as $C \subset A$. An improper subset of A is the set A itself.

Example: The set {All mammals found in Africa} is a proper subset of the set {All animals found in Africa}.

The **complement** A' of a set A consists of all elements of the **universal** set U (all possible elements under consideration) that do not belong to A.

Example: If U = {All letters of the alphabet} and A = {All consonants} then A' = {All vowels}.

A **union** of two sets, written as $A \cup B$, is a set that contains all the elements of both sets. An **intersection** of two sets, written as $A \cap B$, is a set that contains the elements common to both sets.

Example: If A = {2, 4, 6, 8, 10} and B = {1, 2, 3, 4, 5} then $A \cup B$ = {1, 2, 3, 4, 5, 6, 8, 10} and $A \cap B$ = {2, 4}.

Example: Ann, Marie, Tom, Hanna, Mike, Sam and Mitchell are students in a sixth grade class. Ann, Tom and Mike take music. Hanna, Mike, Mitchell and Tom take drama. *A* is the set of children who take music. *B* is the set of children who take drama.
(i) List set *A* and set *B*. (ii) What is the union of *A* and *B*? (iii) What is the intersection of *A* and *B*? (iv) What is the complement of *A*? (v) What is the complement of the intersection of *A* and *B*?

(i) A = {Ann, Tom, Mike};
B = {Hanna, Mike, Mitchell, Tom}
(ii) $A \cup B$ = {Ann, Tom, Mike, Hanna, Mitchell}
(iii) $A \cap B$ = {Tom, Mike}
(iv) A' = {Marie, Hanna, Sam, Mitchell}
(v) ($A \cap B$)' = {Ann, Marie, Hanna, Sam, Mitchell}

Skill 11.2   **Apply addition and multiplication counting principles to determine the number of outcomes related to an event**

Counting Procedures
So far, in all the problems we dealt with, the sample space was given or can be easily obtained. In many real life situations, the sample space and events within it are very large and difficult to find.

There are three techniques to help find the number of elements in one event or a sample space: counting principle, permutations, and combinations.

The Counting Principle: In a sequence of two distinct events in which the first one has n number of outcomes or possibilities, the second one has m number of outcomes or possibilities, the total number or possibilities of the sequence will be

$$n \cdot m$$

Example: A car dealership has three compact models and each model comes in a choice of four colors. How many compact cars are available at the dealership?

Number of available compact cars = (3)(4) = 12

Example: If a license plate consists of three digits followed by three letters, find the possible number of licenses if

    a) the repetition of letters and digits are **not** allowed.

    b) the repetition of letters and digits are allowed.

        a)   Since we have twenty-six letters and ten digits, using the counting principle, we get

        possible # of licenses = (26)(25)(24)(10)(9)(8)
                            = 11,232,000

        b)   Since repetitions are allowed, we get

        possible # of licenses = (26)(26)(26)(10)(10)(10)
                            = 17,576,000

The Addition Principle of Counting states:

If A and B are events, $n(A \, or \, B) = n(A) + n(B) - n(A \cap B)$.

Example: How many ways can you select a black card or a Jack from an ordinary deck of playing cards?

Let B denote the set of black cards and let J denote the set of Jacks. Then,

$$n(B) = 26, n(J) = 4, n(B \cap J) = 2$$
$$= 26 + 4 - 2$$
$$= 28$$

The Addition Principle of Counting for Mutually Exclusive Events states:

If A and B are mutually exclusive events, $n(A \, or \, B) = n(A) + n(B)$.

Example: A travel agency offers 40 possible trips: 14 to Asia, 16 to Europe and 10 to South America. How many trips can be taken to either Asia or Europe through this agency?

Let A denote trips to Asia and let E denote trips to Europe. Then, $A \cap E = \varnothing$ and
$n(A or E) = 14 + 16 = 30$.

Therefore, the number of trips to Asia or Europe is 30.

The Multiplication Principle of Counting for Dependent Events states:

Let A be a set of outcomes of Stage 1 and B a set of outcomes of Stage 2. The number of ways, $n(A and B)$, A and B can occur in a two-stage experiment is given by:

$$n(A and B) = n(A)n(B|A),$$

where $n(B|A)$ denotes the number of ways B can occur given that A has already occurred.

Example: How many ways from an ordinary deck of 52 cards can 2 Jacks be drawn in succession if the first card is drawn but not replaced in the deck and then the second card is drawn?

This is a two-stage experiment for which we wish to compute $n(A and B)$, where A is the set of outcomes for which a Jack is obtained on the first draw and B is the set of outcomes for which a Jack is obtained on the second draw.

If the first card drawn is a Jack, then there are only three remaining Jacks left to choose from on the second draw. Thus, drawing two cards without replacement means the events A and B are dependent.

$$n(A and B) = n(A)n(B|A) = 4 \cdot 3 = 12$$

The Multiplication Principle of Counting for Independent Events states:

Let A be a set of outcomes of Stage 1 and B a set of outcomes of Stage 2. If A and B are independent events, the number of ways, $n(A \text{ and } B)$, A and B can occur in a two-stage experiment is given by:

$$n(A \text{ and } B) = n(A)n(B).$$

Example: How many six-letter combinations can be formed if repetition of letters is not allowed?

A first letter *and* a second letter *and* a third letter *and* a fourth letter *and* a fifth letter *and* a sixth letter must be chosen, therefore there are six stages.

Since repetition is not allowed, there are 26 choices for the first letter, 25 for the second, 24 for the third, 23 for the fourth, 22 for the fifth and 21 for the sixth

n(six-letter combinations without repetition of letters)

$$= 26 \cdot 25 \cdot 24 \cdot 23 \cdot 22 \cdot 21$$
$$= 165,765,600$$

Permutations
In order to understand permutations, the concept of factorials must be discussed.

n factorial, written n!, is represented by n ! = n(n-1)(n-2) . . . . . (2)(1)

$$5! = (5)(4)(3)(2)(1) = 120$$
$$3! = 3(2)(1) = 6$$

By definition: 0! = 1
$\qquad\qquad$ 1! = 1

$$\frac{6!}{6!} = 1 \text{ but } \frac{6!}{2!} \neq 3!$$
$$\frac{6!}{2!} = \frac{6 \cdot 5 \cdot 4 \cdot 3 \cdot 2!}{2!} = 6 \cdot 5 \cdot 4 \cdot 3 = 360$$

The number of permutations represents the number of ways r items can be selected from n items and arranged in a specific order. It is written as $_nP_r$ and is calculated using the following relationship.

$$_nP_r = \frac{n!}{(n-r)!}$$

When calculating permutations, order counts. For example, 2, 3, 4 and 4, 3, 2 are counted as two different permutations. Calculating the number of permutations is not valid with experiments where replacement is allowed.

Example: How many different ways can a president and a vice president be selected from a math class if seven students are available?

We know we are looking for the number of permutations, since the positions of president and vice president are not equal.

$$_7P_2 = \frac{7!}{(7\text{-}2)!} = \frac{7!}{5!} = \frac{7\cdot 6\cdot 5!}{5!} = 7\cdot 6 = 42$$

It is important to recognize that the number of permutations is a special case of the Counting Principle, which can also be used to solve problems dealing with the number of permutations. For instance, in this example we have seven available students to choose a president from. After a president is chosen, we have six available students to choose a vice president. Hence, using the Counting Principle, the ways a president and a vice president can be chosen = 7.6 = 42

Combinations
When dealing with the number of **combinations,** the order in which elements are selected is not important. For instance,

2, 3, 4 and 4, 2, 3 are considered the same combination.

The numbers of combinations represents the number of ways r elements are selected from n elements (in no particular order). The number of combinations is represented by $_nC_r$ and can be calculated using the following relationship.

$$_nC_r = \frac{n!}{(n-r)r!}$$

Example: In how many ways can two students be selected from a class of seven students to represent the class?

Since both representatives have the same position, the order is not important and we are dealing with the number of combinations.

$$_nC_r = \frac{7!}{(7-2)!2!} = \frac{7 \cdot 6 \cdot 5!}{5!2 \cdot 1} = 21$$

Example: In a club there are six women and four men. A committee of two women and one man is to be selected. How many different committees can be selected?

This problem has a sequence of two events. The first event involves selecting two women out of six women and the second event involves selecting one man out of four men. We use the combination relationship to find the number of ways events 1 and 2 can take place and the Counting Principle to find the number of ways the sequence can occur.

Number of possible committees = $_6C_2 \cdot {_4C_1}$

$$\frac{6!}{(6-2)!2!} \times \frac{4!}{(4-1)!1!}$$

$$= \frac{6 \cdot 5 \cdot 4!}{4! \cdot 2 \cdot 1} \times \frac{4 \cdot 3!}{3! \cdot 1}$$

$$= (15) \times (4) = 60$$

Skill 11.3 **Determine probabilities of simple and compound events (e.g., dependent, independent, mutually exclusive, conditional)**

In probability, the **sample space** is a list of all possible outcomes of an experiment. For example, the sample space of tossing two coins is the set {HH, HT, TT, TH} where heads is H and tails is T, and the sample space of rolling a six-sided die is the set {1, 2, 3, 4, 5, 6}.

Probability measures the chances of an event occurring. The probability of an event that must occur, a is **one.** When no outcome is favorable, the probability of an impossible event is **zero.**

P(event) = $\dfrac{\text{number of favorable outcomes}}{\text{number of possible outcomes}}$

If you toss a coin it can land in one of two ways, heads or tails. There are 2 possible outcomes. Since the chance of tossing heads or tails is the same, the outcomes are equally likely.

When a number cube is thrown there are 6 equally likely outcomes. The probability of each outcome is $\frac{1}{6}$.

For Example, when the cube is thrown, the probability of its landing on 5 is $\frac{1}{6}$.

We can write $P(5) = \frac{1}{6}$

If an event has "*n*" equally likely outcomes, then the probability of one of the outcomes is $\frac{1}{n}$.

Example: Given one die with faces numbered 1–6, the probability of tossing an even number on one throw is 3/6 or ½ since there are 3 favorable outcomes (even faces) and a total of 6 possible outcomes (faces).

Example: If a fair die is rolled,
a) Find the probability of rolling an odd number
b) Find the probability of rolling a number less than three.

a) The sample space is

S = {1, 2, 3, 4, 5, 6} and the event representing odd numbers is

E = {1, 3, 5}

Hence, the probability of rolling an odd number is

$$P(E) = \frac{n(E)}{n(S)} = \frac{3}{6} = \frac{1}{2} \text{ or } 0.5$$

b) The event of rolling a number less than three is represented by

A = {1, 2}

Hence, the probability of rolling a number less than three is

$$P(A) = \frac{n(A)}{n(S)} = \frac{2}{6} = \frac{1}{3} \text{ or } 0.33$$

Example: A class has 30 students. Out of the thirty students, 24 are male. Assuming all the students have the same chance of being selected, find the probability of selecting a female. (Only one person is selected.)

The number of females in the class is
      30 - 24 = 6
Hence, the probability of selecting a female is

$$P(female) = \frac{6}{30} \text{ or } 0.2$$

The previous examples are those of **simple events. Compound events** are comprised of more than one simple event and the combined probability of the events occurring depends on the relationship between the events.

If A and B are **independent** events, the outcome of event A does not affect the outcome of event B or vice versa. In this case, the multiplication rule is used to find joint probability, i.e. the probability of both events occurring.

$$P(A \text{ and } B) = P(A) \times P(B)$$

Example: The probability that a patient is allergic to aspirin is 0.30. If the probability of a patient having a window in his/her room is 0.40, find the probability that the patient is allergic to aspirin **and** has a window in his/her room.

Defining the events: A = The patient being allergic to aspirin.
                     B = The patient has a window in his/her room.

Events A and B are independent, hence
$P(A \text{ and } B) = P(A) \cdot P(B)$
$= (0.30) (0.40)$
$= 0.12$ or 12%

Example: Given a jar containing 10 marbles, 3 red, 5 black, and 2 white, what is the probability of randomly drawing a red marble and then a white marble if the marble is returned to the jar after choosing?

$$3/10 \text{ X } 2/10 = 6/100 = 3/50$$

If the probability of an outcome of a first event is $\frac{a}{m}$, and if the probability of an outcome of a second event is $\frac{b}{n}$ then the probability of the first outcome followed by the second outcome is $\frac{a}{m} \times \frac{b}{n}$

Example:
What is the probability of rolling a 5 on a six-sided die and tossing heads on a coin?

$$\frac{1}{6} \times \frac{1}{2} = \frac{1}{12}$$

Example:
What is the probability of rolling a 5 on each of two dice?

$$\frac{1}{6} \times \frac{1}{6} = \frac{1}{36}$$

When the outcome of the first event affects the outcome of the second event, the events are **dependent**. Any two events that are not independent are dependent. This is also known as **conditional probability.**

Conditional Probability:
Dependent events occur when the probability of the second event depends on the outcome of the first event. For example, consider the two events (A) it is sunny on Saturday and (B) you go to the beach. If you intend to go to the beach on Saturday, rain or shine, then A and B may be independent.

If however, you plan to go to the beach only if it is sunny, then A and B may be dependent. In this situation, the probability of event B will change depending on the outcome of event A.

Suppose you have a pair of dice, one red and one green. If you roll a three on the red die and then roll a four on the green die, we can see that these events do not depend on the other. The total probability of the two independent events can be found by multiplying the separate probabilities.

$$P(A \text{ and } B) = P(A) \times P(B)$$
$$= 1/6 \times 1/6$$
$$= 1/36$$

Many times, events are not independent. Suppose a jar contains 12 red marbles and 8 blue marbles. If you randomly pick a red marble, replace it and then randomly pick again, the probability of picking a red marble the second time remains the same. However, if you pick a red marble, and then pick again without replacing the first red marble, the second pick becomes dependent upon the first pick (conditional probability).

P(Red and Red) with replacement = P(Red) $\times$ P(Red)

$$= 12/20 \times 12/20$$

$$= 9/25$$

P(Red and Red) without replacement = P(Red) $\times$ P(Red)

$$= 12/20 \times 11/19$$

$$= 33/95$$

Example:

What is the probability of picking a Jack of Hearts out of a normal deck of 52 playing cards followed by picking a Queen of Hearts from the same deck of cards after

a) replacing the first card after the first draw card
b) not replacing the first card into the deck after the first draw

a) $$\frac{a}{m} \times \frac{b}{n} = \frac{1}{52} \times \frac{1}{52}$$

$$= \frac{1}{2704}$$

b) $$\frac{1}{52} \times \frac{1}{51} = \frac{1}{2652}$$ Probability of (A and B)

$$= P(A) \times P(B \text{ given } A)$$

Example: Two cards are randomly drawn from a deck of 52 cards without replacement; that is, the first card is not returned to the deck before the second card is drawn. What is the probability of drawing two diamonds?

P(A) = probability of drawing a diamond first
P(B) = probability of drawing a diamond second

P(A) = 13/52 = ¼   P(B) = 12/51 = 4/17

P(A+B) = (¼)(14/17) = 7/34

Example: A class of 10 students has six males and four females. If two students are randomly selected to represent the class, find the probability that

a) the first is a male and the second is a female.
b) the first is a female and the second is a male.
c) both are females.
d) both are males.

Defining the events: F = a female is selected to represent the class.
M = a male is selected to represent the class.

F/M = a female is selected after a male has been selected.

M/F = a male is selected after a female has been selected.

a)   Since F and M are dependent events, it follows that
P(M and F) = P(M) · P(F/M)
$$= \frac{6}{10} \times \frac{4}{9} = \frac{3}{5} \times \frac{4}{9} = \frac{12}{45} = \frac{4}{15}$$

P(F/M) = $\dfrac{4}{9}$ instead of , $\dfrac{4}{10}$ since the selection of a male first changed the sample space from ten to nine students.

b)   P(F and M) = P(F) · P(M/F)

$$= \frac{4}{10} \times \frac{6}{9} = \frac{2}{5} \times \frac{2}{3} = \frac{4}{15}$$

c)   P(F and F) = p(F) · p(F/F)

$$= \frac{4}{10} \times \frac{3}{9} = \frac{2}{5} \times \frac{1}{3} = \frac{2}{15}$$

d)   P(both are males) = p(M and M)

$$= \frac{6}{10} \times \frac{5}{9} = \frac{30}{90} = \frac{1}{3}$$

If events A and B are **mutually exclusive**, i.e. they cannot happen together, the probability of either occurring is given by their sum.

$$P(A \text{ or } B) = P(A) + P(B)$$

Example: A six-sided die is rolled. What is the probability of the die showing a 1 **or** a 3?
Probability of a die showing a 1 = 1/6.
Probability of a die showing a 3 = 1/6.
Thus, the probability of either occurring = 1/6 + 1/6 = 1/3.

If events A and B are not mutually exclusive then

$$P(A \text{ or } B) = P(A) + P(B) - P(A \text{ and } B)$$

Skill 11.4 **Use different graphical representations (e.g., Venn diagrams, tree diagrams) to calculate and interpret probabilities**

Venn diagrams are useful for representing sets of outcomes for events within a sample space. In the diagram shown below, the rectangle represents the sample space while A and B are the sets of outcomes of two mutually exclusive events.

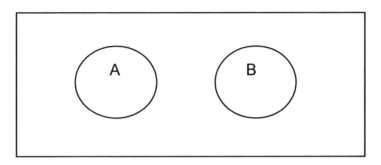

If the events are not mutually exclusive they are represented as shown below. This clearly demonstrates the relationship

$$P(A \text{ or } B) = P(A) + P(B) - P(A \text{ and } B)$$

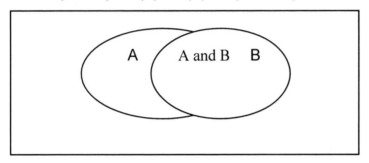

Suppose you want to look at the possible sequence of events for having two children in a family. Since a child will be either a boy or a girl, you would have the following tree diagram to illustrate the possible outcomes:

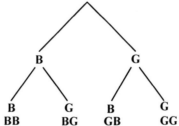

From the diagram, you see that there are 4 possible outcomes, 2 of which are the same.

Example: Make a tree diagram to show all the possible outcomes for rolling a 6-sided number cube and then tossing a coin.

The possible outcomes of rolling a die followed by tossing a coin:

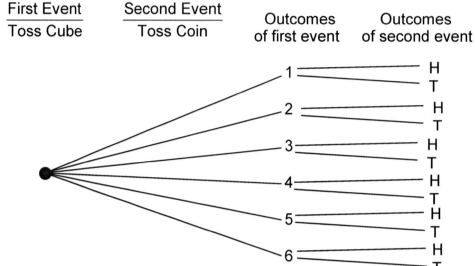

The list of outcomes are: 1H, 1T, 2H, 2T, 3H, 3T, 4H, 4T, 5H, 5T, 6H, 6T.

Using tables

Example: The results of a survey of 47 students are summarized in the table below.

	Black Hair	Blonde Hair	Red Hair	Total
Male	10	8	6	24
Female	6	12	5	23
Total	16	20	11	47

Use the table to answer questions a–c.

a)   If one student is selected at random, find the probability of selecting a male student.

$$\frac{\text{Number of male students}}{\text{Number of students}} = \frac{24}{47}$$

b)   If one student is selected at random, find the probability of selecting a female with red hair.

$$\frac{\text{Number of red hair females}}{\text{Number of students}} = \frac{5}{47}$$

c)   If one student is selected at random, find the probability of selecting a student that does not have red hair.

$$\frac{\text{Red hair students}}{\text{Number of students}} = \frac{11}{47}$$

$$1 - \frac{11}{47} = \frac{36}{47}$$

Skill 11.5   Demonstrate knowledge of methods (e.g., rolling dice, use random numbers) for predicting the probability of an event using trials or simulations

In the preceding sections, we discussed **theoretical probability.** It involves finding the probability of an event by mathematically identifying the sample space and calculating the number of occurrences of an event in that sample space. **Experimental probability**, on the other hand, is based on observation using trials or simulations. The actual number of total occurrences of an event relative to the number of total outcomes is tallied and calculated in order to find the probability.

Example: A coin is tossed 50 times and 27 heads are obtained. What is the experimental probability of flipping heads based on this data? What is the theoretical probability of flipping heads?

Based on the given observation, the experimental probability of flipping heads = 27/50 = 0.54.

Since heads and tails are equally likely in a coin toss, the theoretical probability of getting a head is always ½ or 0.5.

Note: If the coin is tossed many more times, we will find that the experimental probability gets closer to the theoretical probability.

Example: A bag is filled with numbers written on index cards. A number is randomly drawn from the bag, recorded, and replaced. Eric drew a number 12 times. His results are shown below:

He drew the number "4" six times.
He drew the number "1" two times.
He drew the number "3" three times.
He drew the number 9" once.

What is the probability that on his next selection, Eric will draw a "4"? A "1"? A "5"?

There is a 50% chance Eric will draw a "4". $\left(\dfrac{6}{12} = \dfrac{1}{2} = 0.5 = 50\% \right)$

There is a 17% chance Eric will draw a "1". $\left(\dfrac{2}{12} = \dfrac{1}{6} = 0.17 = 17\% \right)$

There is a 0% chance Eric will draw a "5".

SUBAREA V.   MATHEMATICAL PROCESSES AND PERSPECTIVES

COMPETENCY 12.0 UNDERSTAND HOW TO USE A VARIETY OF
REPRESENTATIONS TO COMMUNICATE
MATHEMATICAL IDEAS AND CONCEPTS AND
CONNECTIONS AMONG THEM

Skill 12.1   Communicate mathematical ideas using a variety of
representations (e.g., numeric, tabular, graphical, pictorial,
symbolic)

Examples, illustrations, and symbolic representations are useful
tools in explaining and understanding mathematical concepts. The
ability to create examples and alternative methods of expression
allows students to solve real world problems and better
communicate their thoughts. Many different kinds of graphs,
diagrams, symbols and tables have been used throughout this
guide.

Concrete examples are real world applications of mathematical
concepts. For example, measuring the shadow produced by a tree
or building is a real world application of trigonometric functions;
acceleration or velocity of a car is an application of derivatives; and
finding the volume or area of a swimming pool is a real world
application of geometric principles.

Pictorial illustrations of mathematic concepts help clarify difficult
ideas and simplify problem solving.

Example: Rectangle R represents the 300 students in School A.
Circle P represents the 150 students that participated in band.
Circle Q represents the 170 students that participated in a sport.
70 students participated in both band and a sport.

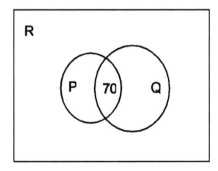

Pictorial representation of above situation.

<u>Example:</u> A marketing company surveyed 200 people and found that 145 people subscribed to at least one magazine, 26 people subscribed to at least one book-of-the-month club, and 19 people subscribed to at least one magazine and one book club. How many people did not subscribe to either a magazine or book club?

Draw a Venn diagram.

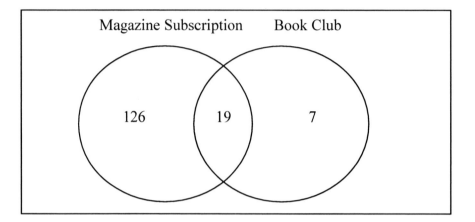

200 − (126 + 19 + 7) = 48 people do not subscribe to either.

Symbolic representation is the basic language of mathematics. Converting data to symbols allows for easy manipulation and problem solving. Students should have the ability to recognize what the symbolic notation represents and convert information into symbolic form. For example, from the graph of a line, students should have the ability to determine the slope and intercepts and derive the line's equation from the observed data. Another possible application of symbolic representation is the formulation of algebraic expressions and relations from data presented in word problem form.

Skill 12.2   **Translate among algebraic, graphical, symbolic, diagrammatic, and other means of presenting mathematical ideas (e.g., set notation, interval notation)**

Mathematical concepts and procedures can take many different forms. Students of mathematics must be able to recognize different forms of equivalent concepts.

For example, we can represent the slope of a line graphically, algebraically, verbally, and numerically. A line drawn on a coordinate plane will show the slope. In the equation of a line, $y = mx + b$, the term m represents the slope. We can define the slope of a line several different ways. The slope of a line is the change in the value of the y divided by the change in the value of x over a given interval. Alternatively, the slope of a line is the ratio of "rise" to "run" between two points. Finally, we can calculate the numeric value of the slope by using the verbal definitions and the algebraic representation of the line.

Most mathematical concepts can be expressed in multiple ways. For example a parabola can be expressed as an equation or a graph. A function can be rewritten as a table. Each way is an equally accurate method of representing the concept, but different techniques may be useful in different situations. It is therefore important to be able to translate any concept into the most appropriate form for addressing any given problem.

$y = (x - 1)^2 - 3$ *is equivalent to*

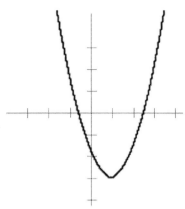

One example of a mathematical idea that can be presented in multiple ways is any group of numbers that is a subset of the real number line. There are three common ways to denote such a set of numbers: graphically on the real number line, in interval notation, and in set notation. For example, the set of numbers consisting of all values greater than negative 2 and less than or equal to 4 could be written in the following three ways:

Graphically:

-4 -3 -2 -1 0 1 2 3 4 5

In this type of notation, open circles exclude an endpoint while closed circles include it.

Interval Notation: (-2, 4]. Interval notation uses round brackets to exclude the endpoint of a set, and square brackets to include it.

Set Notation: $\{-2 < x \le 4\}$

p
Depending on why you need to represent this group of numbers, each form of notation has its advantages. Set notation is the most flexible, since finite sets can be written as lists within the brackets. However, it is often far more cumbersome than interval notation, and in some circumstances graphical notation may be the most clear.

Skill 12.3   Translate between mathematical language, notation, and symbols and everyday language

An algebraic formula is an equation that describes a relationship among variables. While it is not often necessary to derive the formula, one must know how to rewrite a given formula in terms of a desired variable.

Example: Given that the relationship of voltage, V, applied across a material with electrical resistance, R, when a current, I, is flowing through the material is given by the formula $V = IR$. Find the resistance of the material when a current of 10 milliamps is flowing, when the applied voltage is 2 volts.

$V = IR$. Solve for R.
$IR = V$; $R = V/I$ Divide both sides by I.
When $V = 2$ volts; $I = 10 \times 10^{-3}$ amps;

$$R = \frac{2}{10^1 \times 10^{-3}}$$

$$R = \frac{2}{10^{-2}}$$ Substituting in $R = V/I$, we get,

$$R = 2 \times 10^2$$

$$R = 200 \text{ ohms}$$

Another example of translating between mathematical language and everyday language is the conversion of recipes to different serving sizes. The conversion factor, the number we multiply each ingredient by, is:

$$\text{Conversion Factor} = \frac{\text{Number of Servings Needed}}{\text{Number of Servings in Recipe}}$$

Example: Consider the following recipe.

3 cups flour
½ tsp. baking powder
2/3 cups butter
2 cups sugar
2 eggs

If the above recipe serves 8, how much of each ingredient do we need to serve only 4 people?

First, determine the conversion factor.

$$\text{Conversion Factor} = \frac{4}{8} = \frac{1}{2}$$

Next, multiply each ingredient by the conversion factor.

3 x ½ =	1 ½ cups flour
½ x ½ =	¼ tsp. baking powder
2/3 x ½ = 2/6 =	1/3 cups butter
2 x ½ =	1 cup sugar
2 x ½ =	1 egg

Skill 12.4   Use appropriate mathematical terminology with precision and accuracy

Unlike in ordinary conversation, language in mathematics has very precise and specific meaning. This precision is required so that mathematical ideas expressed in words can be converted unambiguously into symbols and vice versa.

Since a lot of mathematical terminology is comprised of words that are used ordinarily as well (e.g. variable or slope), it is critical that students understand the importance of the **accurate use of terms**. The best way for a teacher to achieve this is to model precision in communication and to insist that students themselves do so when communicating in mathematics class verbally or in writing.

Apart from the accurate use of individual terms, one must also use **completeness** in every mathematical statement made. This means that every mathematical statement must include necessary qualifications and conditions so that there is no doubt as to its meaning.

Skill 12.5   Analyze and evaluate a mathematical model (e.g. equation, graph, pictorial representation) in terms of its appropriateness and usefulness in a given situation

See Skill 12.1.

Skill 12.6 **Recognize connections among different concepts and areas of mathematics (e.g., algebra, geometry) and use them to solve problems**

Mathematical concepts, and the skills needed to apply those concepts, overlap into different areas of mathematics. This is especially visible when algebra is needed to solve geometric problems. For example, you are asked to find the area of a rectangle, but one of the sides is labeled as a variable. You will then need to use algebra to solve for the variable.

Example: A construction worker on a scaffold 50 ft above the ground dropped a hammer to another worker 36 feet below him. The height of the hammer over time is represented by the function $h = -16x^2 + 50$.

This is an example of a problem with connections among different concepts. The path of the hammer can be shown graphically as well as algebraically. The problem also connects math with the science of gravity ($h = -16x^2$).

The height of the hammer as a graph:

Height of Hammer

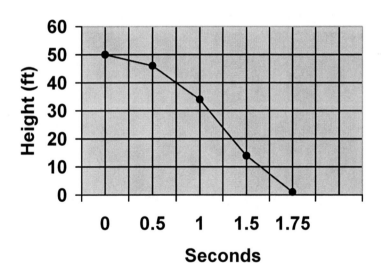

The height of the hammer algebraically:

$h = -16x^2 + 50$

$h = -16(0)^2 + 50$

$h = 50$

The height of the hammer in a table:

x	y
0	50
0.5	46
1	34
1.5	14
1.75	1

Example: A family wants to enclose 3 sides of a rectangular garden with 200 feet of fence. In order to have a garden with an area of **at least** 4,800 square feet, find the dimensions of the garden. Assume that a wall or a fence already borders the fourth side of the garden

Existing Wall

Let x = distance
from the wall

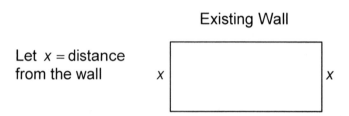

Then $2x$ feet of fence is used for these 2 sides. The side opposite the existing wall would use the remainder of the 200 feet of fence, that is, $200 - 2x$ feet of fence. Therefore the width (w) of the garden is x feet and the length (l) is $200 - 2x$ feet.

The area is calculated using the formula $A = lw$.

$A = x(200 - 2x) = 200x - 2x^2$ which must be greater than or equal to 4,800 sq. ft. yielding the inequality $4800 \leq 200x - 2x^2$. Subtract 4,800 from each side and the inequality becomes $2x^2 - 200x + 4800 \leq 0$ and can be solved for x.

$$200x - 2x^2 \geq 4800$$
$$-2x^2 + 200x - 4800 \geq 0$$
$$2\left(-x^2 + 100x - 2400\right) \geq 0$$
$$-x^2 + 100x - 2400 \geq 0$$
$$(-x + 60)(x - 40) \geq 0$$
$$-x + 60 \geq 0$$
$$-x \geq -60$$
$$x \leq 60$$
$$x - 40 \geq 0$$
$$x \geq 40$$

So the area will be at least 4,800 square feet if the width of the garden is from 40 up to 60 feet. (The length of the rectangle would vary from 120 feet to 80 feet depending on the width of the garden.)

Skill 12.7   **Identify and apply mathematics in contexts outside of mathematics**

Artists, musicians, scientists, social scientists, and business people use mathematical modeling to solve problems in their disciplines. These disciplines rely on the tools and symbols of mathematics to model natural events and manipulate data. Mathematics is a key aspect of visual art.

Artists use the geometric properties of shapes, ratios, and proportions in creating paintings and sculptures. For example, mathematics is essential to the concept of perspective. Artists must determine the appropriate lengths and heights of objects to portray three-dimensional distance in two dimensions.

Mathematics is also an important part of music. Many musical terms have mathematical connections. For example, the musical octave contains twelve notes and spans a factor of two in frequency. In other words, the frequency, the speed of vibration that determines tone and sound quality, doubles from the first note in an octave to the last. Thus, starting from any note we can determine the frequency of any other note with the following formula.

$$Freq = note \times 2^{N/12}$$

Where N is the number of notes from the starting point and note is the frequency of the starting note. Mathematical understanding of frequency plays an important role in the tuning of musical instruments.

In addition to the visual and auditory arts, mathematics is an integral part of most scientific disciplines. The uses of mathematics in science are almost endless. The following are but a few examples of how scientists use mathematics. Physical scientists use vectors, functions, derivatives, and integrals to describe and model the movement of objects. Biologists and ecologists use mathematics to model ecosystems and study DNA. Finally, chemists use mathematics to study the interaction of molecules and to determine proper amounts and proportions of reactants.

Many social science disciplines use mathematics to model and solve problems. Economists, for example, use functions, graphs, and matrices to model the activities of producers, consumers, and firms. Political scientists use mathematics to model the behavior and opinions of the electorate. Finally, sociologists use mathematical functions to model the behavior of humans and human populations.

Finally, mathematical problem solving and modeling is essential to business planning and execution. For example, businesses rely on mathematical projections to plan business strategy. Additionally, stock market analysis and accounting rely on mathematical concepts.

COMPETENCY 13.0 UNDERSTAND MATHEMATICAL REASONING, PROOF, AND PROBLEM-SOLVING STRATEGIES IN MATHEMATICS AND OTHER CONTEXTS

Skill 13.1   Demonstrate knowledge of the nature of proof in mathematics, including direct and indirect proof, and the use of counterexamples

Direct Proofs
A direct proof proceeds systematically step by step from given information to a certain conclusion. Each step must be justified with reasons. In a two-column proof in geometry, the left side of the proof should be the given information, or statements that could be proved by deductive reasoning. The right column of the proof consists of the reasons used to determine that each statement to the left was verifiably true. The right side can identify given information, or state theorems, postulates, definitions or algebraic properties used to prove that particular line of the proof is true. This is an example of deductive reasoning. Both inductive and deductive reasoning are discussed in the next section in Skill 13.2.

Indirect Proofs
An indirect proof assumes that the opposite of the conclusion is true and then shows that this assumption will lead to a contradiction. Keeping the hypothesis and given information the same, proceed to develop the steps of the proof, looking for a statement that contradicts your original assumption or some other known fact. This contradiction indicates that the assumption you made at the beginning of the proof was incorrect; therefore, the original conclusion has to be true.

A **counterexample** is an exception to a proposed rule or conjecture that disproves the conjecture. For example, the existence of a single non-brown dog disproves the conjecture "all dogs are brown". Thus, any non-brown dog is a counterexample.

In searching for mathematic counterexamples, one should consider extreme cases near the ends of the domain of an experiment and special cases where an additional property is introduced.

Examples of extreme cases are numbers near zero and obtuse triangles that are nearly flat. An example of a special case for a problem involving rectangles is a square because a square is a rectangle with the additional property of symmetry.

Example: Identify a counterexample for the following conjectures.

1. If n is an even number, then $n + 1$ is divisible by 3.

$n = 4$
$n + 1 = 4 + 1 = 5$
5 is not divisible by 3.

2. If n is divisible by 3, then $n^2 - 1$ is divisible by 4.

$n = 6$
$n^2 - 1 = 6^2 - 1 = 35$
35 is not divisible by 4.

Skill 13.2   **Apply correct mathematical reasoning to draw valid conclusions and to construct and evaluate arguments and proofs**

See Skill 13.3.

Skill 13.3   **Identify inductive and deductive reasoning and use them to develop and investigate the validity of conjectures**

Inductive reasoning is the process of finding a pattern from a group of examples. That pattern is the conclusion that this set of examples seemed to indicate. It may be a correct conclusion or it may be an incorrect conclusion because other examples may not follow the predicted pattern.

Example:

Suppose:
On Monday, Mr. Peterson eats breakfast at Morton's.
On Tuesday, Mr. Peterson eats breakfast at Morton's.
On Wednesday, Mr. Peterson eats breakfast at Morton's.
On Thursday, Mr. Peterson eats breakfast at Morton's again.

Conclusion: On Friday, Mr. Peterson will eat breakfast at Morton's again.

This is a conclusion based on inductive reasoning. Based on several days' observations, you conclude that Mr. Peterson will eat at Morton's. This may or may not be true, but it is a conclusion arrived at by inductive thinking.

A **simple statement** represents a simple idea, that can be described as either "true" or "false", but not both. A simple statement is represented by a small letter of the alphabet.

For instance, "Today is Monday" is a simple statement that can be described as either true or false. We can write p = "Today is Monday". On the other hand, "John, please be quite" is not considered a simple statement in our study of logic, since we cannot assign a truth value to it.

Simple statements joined together by **connectives** ("and", "or", "not", "if then", and "if and only if") result in **compound statements**. Note that compound statements can also be formed using "but", "however", or "never the less". A compound statement can be assigned a truth value.

Conditional statements are frequently written in "if-then" form. The "if" clause of the conditional is known as the **hypothesis**, and the "then" clause is called the **conclusion**. In a proof, the hypothesis is the information that is assumed to be true, while the conclusion is what is to be proven true. A conditional is considered to be of the form: **If p, then q** where p is the hypothesis and q is the conclusion.

p → q is read "if p then q".
~ (statement) is read "it is not true that (statement)".

Quantifiers are words describing a quantity under discussion. These include words like "all', "none" (or "no"), and "some".

If a statement is true, then its **negation** must be false (and vice versa).

A Summary of Negation Rules:

statement	negation
(1) q	(1) <u>not</u> q
(2) <u>not</u> q	(2) q
(3) π <u>and</u> s	(3) (not π) <u>or</u> (not s)
(4) π <u>or</u> s	(4) (not π) <u>and</u> (not s)
(5) if p, then q	(5) (p) <u>and</u> (not q)

Example: Select the statement that is the negation of "some winter nights are cold".

A. All winter nights are not cold.
B. Some winter nights are cold.
C. All winter nights are cold.
D. None of the winter nights are cold.

Negation of "some are" is "none are". So the negation statement is "none of the winter nights are cold". So the answer is D.

Example: Select the statement that is the negation of "if it rains, then the beach party will not be held".

A. If it does not rain, then the beach party will be held.
B. If the beach party is held, then it will not rain.
C. It does not rain and the beach party will be held.
D. It rains and the beach party will be held.

Negation of "if p, then q" is "p and (not q)". So the negation of the given statement is "it rains and the beach party will be held". So select D.

Example: Select the negation of the statement "If they get elected, then all politicians go back on election promises".

A. If they get elected, then many politicians go back on election promises.
B. They get elected and some politicians go back on election promises.
C. If they do not get elected, some politicians do not go back on election promises.
D. None of the above statements is the negation of the given statement.

Identify the key words of "if...then" and "all...go back". The negation of the given statement is "they get elected and none of the politicians go back on election promises". So select response D, since A, B, and C, statements are not the negations.

Example: Select the statement that is the negation of "the sun is shining bright and I feel great".

A. If the sun is not shining bright. I do not feel great.
B. The sun is not shining bright and I do not feel great.
C. The sun is not shining bring or I do not feel great.
D. the sun is shining bright and I do not feel great.

The negation of "r and s" is "(not r) or (not s)". So the negation of the given statement is "the sun is not shining bright or I do not feel great". We select response C.

Conditional statements can be diagrammed using a **Venn diagram**. A diagram can be drawn with one circle inside another circle. The inner circle represents the hypothesis. The outer circle represents the conclusion. If the hypothesis is taken to be true, then you are located inside the inner circle. If you are located in the inner circle then you are also inside the outer circle, so that proves the conclusion is true.

Example: If you are in Pittsburgh, then you are in Pennsylvania.

In this statement "you are in Pittsburgh" is the hypothesis.
In this statement "you are in Pennsylvania" is the conclusion.

Example: If an angle has a measure of 90 degrees, then it is a right angle.

In this statement "an angle has a measure of 90 degrees" is the hypothesis. In this statement "it is a right angle" is the conclusion.

Deductive reasoning is the process of arriving at a conclusion based on other statements that are all known to be true, such as theorems, axioms, or postulates. Conclusions found by deductive reasoning based on true statements will **always** be true.

A symbolic argument consists of a set of premises and a conclusion in the format of of if [Premise 1 and premise 2] then [conclusion].

An argument is **valid** when the conclusion follows necessarily from the premises. An argument is **invalid** or a fallacy when the conclusion does not follow from the premises.

There are 4 standard forms of valid arguments which must be remembered.

1. Law of Detachment	If p, then q p, Therefore, q	(premise 1) (premise 2)
2. Law of Contraposition	If p, then q not q, Therefore not p	
3. Law of Syllogism	If p, then q If q, then r Therefore if p, then r	
4. Disjunctive Syllogism	p or q not p Therefore, q	

<u>Example</u>: Can a conclusion be reached from these two statements?

A.   All swimmers are athletes.
        All athletes are scholars.

In "if-then" form, these would be:
    If you are a swimmer, then you are an athlete.
    If you are an athlete, then you are a scholar.

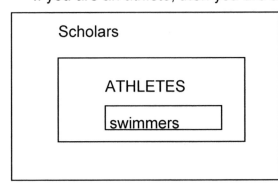

Clearly, if you are a swimmer, then you are also an athlete. This includes you in the group of scholars.

B.   All swimmers are athletes.
        All wrestlers are athletes.

In "if-then" form, these would be:
    If you are a swimmer, then you are an athlete.
    If you are a wrestler, then you are an athlete.

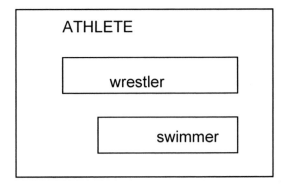

Clearly, if you are a swimmer or a wrestler, then you are also an athlete. This does NOT allow you to come to any other conclusions.

A swimmer may or may NOT also be a wrestler. Therefore, NO CONCLUSION IS POSSIBLE.

Example: Determine whether statement A, B, C, or D can be deduced from the following:

(i) If John drives the big truck, then the shipment will be delivered.

(ii) The shipment will not be delivered.

A. John does not drive the big truck.
B. John drives the big truck.
C. The shipment will not be delivered.
D. None of the above conclusion is true.

Let p: John drives the big truck.
    q: The shipment is delivered.

statement (i) gives $p \rightarrow q$, statement (ii) gives $\sim q$. This is the Law of Contraposition.

Therefore, the logical conclusion is $\sim p$ or "John does not drive the big truck". So the answer is response A.

Example: Given that:
(i) Peter is a Jet Pilot or Peter is a Navigator.
(ii) Peter is not a Jet Pilot

Determine which conclusion can be logically deduced.

A. Peter is not a Navigator.
B. Peter is a Navigator.
C. Peter is neither a Jet Pilot nor a Navigator.
D. None of the above is true.

Let p: Peter is a Jet Pilot
    q: Peter is a Navigator.

We have $p \vee q$ from statement (i)
       $\sim p$ from statement (ii)

So choose response B.

Try These:

What conclusion, if any, can be reached? Assume each statement is true, regardless of any personal beliefs.

1. If the Red Sox make it to the World Series, I will buy World Series tickets. I bought World Series tickets.

2. If an angle's measure is between 0° and 90°, then the angle is acute. Angle B is not acute.

3. Students who do well in geometry will succeed in college. Annie is doing extremely well in geometry.

4. Left-handed people are witty and charming. You are left-handed.

Answers:

Question #1	The Red Sox won the World Series.
Question #2	Angle B is not between 0 and 90 degrees.
Question #3	Annie will do well in college.
Question #4	You are witty and charming.

Skill 13.4 **Select an appropriate problem-solving strategy (e.g., estimation, working backward, drawing a diagram) for investigating or solving a particular problem**

Successful math teachers introduce their students to multiple problem-solving strategies and create a classroom environment where free thought and experimentation are encouraged.

Teachers can promote problem solving by allowing multiple attempts at problems, giving credit for reworking test or homework problems, and encouraging the sharing of ideas through class discussion. There are several specific problem-solving skills with which teachers should be familiar.

The **guess-and-check** strategy calls for students to make an initial guess at the solution, check the answer, and use the outcome as a guide for the next guess. With each successive guess, the student should get closer to the correct answer. Constructing a table from the guesses can help organize the data.

Example: There are 100 coins in a jar. 10 are dimes. The rest are pennies and nickels. There are twice as many pennies as nickels. How many pennies and nickels are in the jar?

There are 90 total nickels and pennies in the jar (100 coins – 10 dimes).

There are twice as many pennies as nickels. Make guesses that fulfill the criteria and adjust based on the answer found. Continue until we find the correct answer, 60 pennies and 30 nickels.

Number of Pennies	Number of Nickels	Total Number of Pennies and Nickels
40	20	60
80	40	120
70	35	105
60	30	90

When solving a problem where the final result and the steps to reach the result are given, students must **work backwards** to determine what the starting point must have been.

Example: John subtracted seven from his age, and divided the result by 3. The final result was 4. What is John's age?

Work backward by reversing the operations.
$4 \times 3 = 12$;
$12 + 7 = 19$
John is 19 years old.

Estimation and testing for **reasonableness** are related skills students should employ both before and after solving a problem. These skills are particularly important when students use calculators to find answers.

Example: Find the sum of 4,387 + 7,226 + 5,893.

4,400 + 7,200 + 5,900 = 17,500	Estimation.
4,387 + 7,226 + 5,893 = 17,506	Actual sum.

By comparing the estimate to the actual sum, students can determine that their answer is reasonable.

Skill 13.5 **Apply mathematical concepts, strategies, and appropriate technologies across the curriculum and in everyday contexts to model and solve problems**

Teachers can increase student interest in math and promote learning and understanding by relating mathematical concepts to the lives of students. Instead of using only abstract presentations and examples, teachers should relate concepts to real-world situations to shift the emphasis from memorization and abstract application to understanding and applied problem solving. In addition, relating math to careers and professions helps illustrate the relevance of math and aids in the career exploration process.

For example, when teaching a unit on the geometry of certain shapes, teachers can ask students to design a structure of interest to the student using the shapes in question. This exercise serves the dual purpose of teaching students to learn and apply the properties (e.g. area, volume) of shapes while demonstrating the relevance of geometry to architectural and engineering professions.

Finally, there are many forms of **technology** available to math teachers. For example, students can test their understanding of math concepts by working on skill specific computer programs and websites. Graphing calculators can help students visualize the graphs of functions. Teachers can also enhance their lectures and classroom presentations by creating multimedia presentations.

Technology is increasingly important in schools today. Knowing what each item is, how to use each item and the importance of each item beyond the walls of the school is extremely important for teachers and students. We will cover some of the most important technology tools below:

Computers: Personal computing, word processing, research, and multi-media devices that utilize hard drive or "memory" in order to save, transform, and process information. It is easy to look past the importance of the computer, itself, in the wake of internet technology. However, many children from low socioeconomic backgrounds throughout the country lack access to computers and may not have basic computing skills. Indeed, many teachers must be taught to utilize some of the more advanced features that could help them with their daily work. In general, the computer is viewed as a tool for storage and efficient processing.

Internet: Platform for the sharing and organization of information throughout the world. It began with simple phone connections between dozens of computers in the 70s and 80s. In the 90s, it became a standard feature on virtually every computer. Through websites, the internet allows people to view, post, and find anything that is available for public viewing. This is helpful when it comes to classroom research; however, there is plenty of material that is inappropriate for children and irrelevant to school. Care should be given to filtering out websites that are inappropriate for students.

Video Projection: Video in the classroom is a common tool to enhance the learning of students. Indeed, video has become a "text" in itself, much like literature. While the internet has been a good source for information, video continues to be a great source of refined, carefully structured information for teachers and students. In addition, it is a great way for students to express themselves, as they develop and present multi-media presentations, for example.

CDs or DVDs: Data discs that are inserted into computers. Usually, CDs and DVDs are discs contain computer programs. It used to be that they contained information, such as encyclopedic information. Now, that material is typically stored on a computer hard drives or servers.

While computers cannot replace teachers, they can be used to enhance the curriculum. Computers may be used to help students practice basic skills. Many excellent programs exist to encourage higher-order thinking skills, creativity and problem solving. Learning to use technology appropriately is an important preparation for adulthood. Computers can also show the connections between mathematics and the real world.

Calculators: Handheld mathematical computation devices. Usually, these help students and teachers perform basic mathematical functions. More advanced functions can be performed by various computer programs.

ESSENTIAL TIPS FOR EVERY MATH TEACHER

Pedagogical principles and teaching methods are important for all teachers. They are particularly critical, though, for math teaching since math teachers not only face the difficulty of communicating the subject matter to students but also that of surmounting an all-pervasive cultural fear of mathematics. Math teachers need to take particular care to foster learning in a non-threatening environment that is at the same time genuinely stimulating and challenging.

The National Council of Teachers of Mathematics (NCTM) (http://www.nctm.org/) Principles and Standards emphasizes the teacher's obligation to support all students not only in developing basic mathematics knowledge and skills but also in their ability to understand and reason mathematically to solve problems relevant to today's world. The use of technology in the classroom is strongly advocated.

Resources for middle school teachers are available on the NCTM website at http://www.nctm.org/resources/middle.aspx.

The Mathematics Pathway (http://msteacher.org/math.aspx) on the National Science Digital Library (NSDL) Middle School Portal provides a very comprehensive and rich treasure trove of helpful material linking to various resources on the web including articles as well as interactive instructional modules on various topics.

The Drexel University Math Forum website provides the opportunity to interact with mentors and other math educators online. Some of the material on this website requires paid subscription but there are openly available archives as well. An overview of what the site provides is available at http://mathforum.org/about.forum.html. You may find the "Teacher2Teacher" service particularly useful; you can ask questions or browse the archives for a wealth of nitty-gritty everyday teaching information, suggestions and links to teaching tools.

This website for sixth grade contains animated lessons, discussions of strategies and a glossary of terms using few words and plenty of illustrations.
http://students.resa.net/stoutcomputerclass/1math.htm

Other instructional and professional development resources:
http://archives.math.utk.edu/k12.html
http://www.learnalberta.ca/Launch.aspx?content=/content/mesg/html/math6web/math6shell.html
http://mmap.wested.org/webmath/

Pedagogical Principles

Maintain a supportive, non-threatening environment
Many students unfortunately perceive mathematics as a threat. This becomes a particular critical issue at the middle school level where they learn algebra for the first time and are required to think in new ways. Since fear "freezes" the brain and makes thinking really difficult, a student's belief that he is no good at math becomes a self-fulfilling prophecy. A teacher's primary task in this situation is to foster a learning environment where every student feels that he or she can learn to think mathematically. Here are some ways to go about this:

Accept all comments and questions: Acknowledge all questions and comments that students make. If what the student says is inaccurate or irrelevant to the topic in hand, point that out gently but also show your understanding of the thought process that led to the comment. This will encourage students to speak up in class and enhance their learning.

Set aside time for group work: Assign activities to groups of students comprised of mixed ability levels. It is often easier for students to put forward their own thoughts as part of a friendly group discussion than when they are sitting alone at their desks with a worksheet. The more proficient students can help the less able ones and at the same time clarify their own thinking. You will essentially be using the advanced students in the class as a resource in a manner that also contributes to their own development. The struggling students will feel supported by their peers and not isolated from them.

Encourage classroom discussion of math topics: For instance, let the whole class share different ways in which they approach a certain problem. It will give you insight into your students' ways of thinking and make it easier to help them. It will allow even those who just listen to understand and correct errors in their thinking without being put on the spot.

Engage and challenge students
Maintaining a non-threatening environment should not mean dumbing down the math content in the classroom. The right level of challenge and relevance to their daily lives can help to keep students interested and learning. Here are some ideas:

Show connections to the real world: Use real life examples of math problems in your teaching. Some suggestions are given in the next section. Explain the importance of math literacy in society and the pitfalls of not being mathematically aware. An excellent reference is "The 10 Things All Future Mathematicians and Scientists Must Know" by Edward Zaccaro. The title of the book is misleading since it deals with things that every educated person, not just mathematicians and scientists, should know.

Use technology: Make use of calculators and computers including various online, interactive resources in your teaching. The natural affinity today's children have for these devices will definitely help them to engage more deeply in their math learning.

Demonstrate "messy" math: Children often have the mistaken belief that every math problem can be solved by following a particular set of rules; they either know the rules or they don't. In real life, however, math problems can be approached in different ways and often one has to negotiate several blind alleys before getting to the real solution. Children instinctively realize this while doing puzzles or playing games. They just don't associate this kind of thinking with classroom math. The most important insight any math teacher can convey to students is the realization that even if they don't know how to do a problem at first, they can think about it and figure it out as long as they are willing to stay with the problem and make mistakes in the process. An obvious way to do this, of course, is to introduce mathematical puzzles and games in the classroom. The best way, however, is for teachers themselves to take risks occasionally with unfamiliar problems and demonstrate to the class how one can work one's way out of a clueless state.

Show the reasoning behind rules: Even when it is not a required part of the curriculum, explain, whenever possible, how a mathematical rule is derived or how it is connected to other rules. For instance, in explaining the rule for finding the area of a trapezoid, show how one can get to it by thinking of the trapezoid as two triangles. This will reinforce the students' sense of mathematics as something that can be logically arrived at and not something for which they have to remember innumerable rules. Another way to reinforce this idea is to do the same problem using different approaches.

Be willing to take occasional side trips: Be flexible at times and go off topic in order to explore more deeply any questions or comments from the students. Grab a teaching opportunity even if it is irrelevant to the topic under discussion.

Help every student gain a firm grasp of fundamentals
While discussion, reasoning and divergent thinking is to be encouraged, it can only be done on a firm scaffolding of basic math knowledge. A firm grasp of math principles, for most people, does require rote exercises and doing more and more of the same problems. Just as practicing scales is essential for musical creativity, math creativity can only be built on a foundation strengthened by drilling and repetition. Many educators see independent reasoning and traditional rule-based drilling as opposing approaches. An effective teacher, however, must maintain a balance between the two and ensure that students have the basic tools they need to think independently.

Make sure all students actually know basic math rules and concepts: Test students regularly for basic math knowledge and provide reinforcement with additional practice wherever necessary.

<u>Keep reviewing old material</u>: Don't underestimate your students' ability to forget what they haven't seen in a while. Link new topics whenever possible with things your students have learned before and take the opportunity to review previous learning. Most math textbooks nowadays have a spiral review section created with this end in mind.

<u>Keep mental math muscles strong:</u> The calculator, without question, is a very valuable learning tool. Many students, unfortunately, use it as a crutch to the point that they lose the natural feel for numbers and ability to estimate that people develop through mental calculations. As a result, they are often unable to tell when they punch a wrong button and get a totally unreasonable answer. Take your students through frequent mental calculation exercises; you can easily integrate it into class discussions. Teach them useful strategies for making mental estimates.

Specific Teaching Methods

Some commonly used teaching techniques and tools are described below along with links to further information. The links provided in the first part of this chapter also provide a wealth of instructional ideas and material.

A very useful resource is the book "Family Math: The Middle School Years" from the Lawrence Hall of Science, University of California at Berkeley. Although this book was developed for use by families, teachers in school can choose from the many simple activities and games used to reinforce two significant middle school skills, algebraic reasoning and number sense. A further advantage is that all the activities are based on NCTM standards and each activity lists the specific math concepts that are covered.

Here are some tools you can use to make your teaching more effective:

Classroom openers
To start off your class with stimulated, interested and focused students, provide a short opening activity everyday. You can make use of thought-provoking questions, puzzles or tricks. Also use relevant puzzles or tricks to illustrate specific topics at any point in your class. The following website provides some ideas:
http://mathforum.org/k12/k12puzzles/

Real life examples

Connect math to other aspects of your students' lives by using examples and data from the real world whenever possible. It will not only keep them engaged, it will also help answer the perennial question "Why do we have to learn math?" Online resources to get you started:

1. Using weather concepts to teach math:
   http://www.nssl.noaa.gov/edu/ideas/

2. Election math in the classroom:
   http://mathforum.org/t2t/faq/election.html

3. Math worksheets related to the Iditarod, an annual Alaskan sled dog race:
   http://www.educationworld.com/a_lesson/lesson/lesson302.shtml

4. Personal finance examples:
   http://www.publicdebt.treas.gov/mar/marmoneymath.htm

5. Graphing with real data:
   http://www.middleweb.com/Graphing.html

Manipulatives

Manipulatives can help all students learn; particularly those oriented more towards visual and kinesthetic learning. Here are some ideas for the use of manipulatives in the classroom:

1. Use tiles, pattern blocks or geoboards to demonstrate geometry concepts such as shapes, area and perimeter. In the example shown below, 12 tiles are used to form different rectangles.

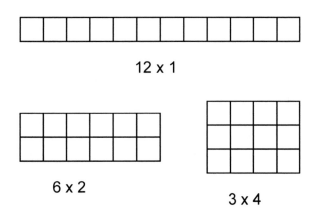

12 x 1

6 x 2

3 x 4

2. Stacks of blocks representing numbers are useful for teaching basic statistics concepts such as mean, median and mode. Rearranging the blocks to make each stack the same height would demonstrate the mean or average value of the data set. The example below shows a data set represented by stacks of blocks. Rearranging the blocks to make the height of each stack equal to three shows that this is the mean value.

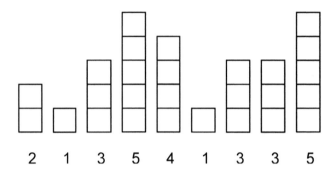

2 1 3 5 4 1 3 3 5

3. Tiles, blocks, or other countable manipulatives such as beans can also be used to demonstrate numbers in different bases. Each stack will represent a place with the number of blocks in the stack showing the place value.

4. Playing cards can be used for a discussion of probability.

5. Addition and subtraction of integers, positive and negative, is a major stumbling block for many middle school students. Two sets of tiles, marked with pluses and minuses respectively, can be used to demonstrate these concepts visually with each "plus" tile canceling a "minus" tile.

| + | + | + | + | | - | - | - | - | - | +4-5=-1

| - | - | - | | - | - | - | - | -3-4=-7

6. Percentages may be visualized using two parallel number lines, one showing the actual numbers, the other showing the percentages.

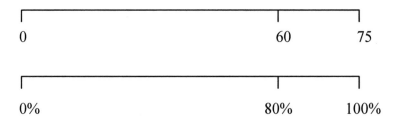

A practical demonstration of percent changes can be made by photocopying a figure using different copier magnifications.

7. Algeblocks are blocks designed specifically for the teaching of algebra with manipulatives:
http://www.etacuisenaire.com/algeblocks/algeblocks.jsp

Software
Many of the online references in this section link to software for learning. A good site that provides easy to use virtual manipulatives as well as accompanying worksheets in some cases is the following:
http://boston.k12.ma.us/teach/technology/select/index.html

Spreadsheets can be very effective math learning tools. Here are some ideas for using spreadsheets in the classroom:
http://www.angelfire.com/wi2/spreadsheet/necc.html

Word problem strategies
Word problems, a challenge for many students even in elementary school, become more complicated and sometimes intimidating in the middle grades. Here are some ideas students can use to tackle them:

1. Identify significant words and numbers in the problem. Highlight or underline them. If necessary, write them in the form of a table.

2. Draw diagrams to clarify the problem. Put down the main items or events and numbers on the diagram and show the relationships between them.

3. Rewrite the problem using fewer and simpler words. One way is to have a standard format for this as shown in the example below.
Problem: Calculate the cost of 3 pencils given that 5 pencils cost 25 cents.
Rewrite as:
Cost of 5 pencils = 25 cents
Cost of 1 pencil = 25/5 = 5 cents
Cost of 3 pencils = 5 X 3 = 15 cents

4. If you have no idea how to approach the problem, try the guess and check approach at first. That will give you a sense of the kind of problem you are dealing with.

5. Create similar word problems of your own.

Equation rule

Solving algebraic equations is a challenge for many learners particularly when they think they need to remember many different rules. Emphasize the fact that they only need to keep only one rule in mind whether they are adding, subtracting, multiplying or dividing numbers or variables:

"Do the same thing to both sides"

A balance or teeter-totter metaphor can help to clarify their understanding of equations. You can also use manipulatives to demonstrate.

Mental math practice

Give students regular practice in doing mental math. The following website offers many mental calculation tips and strategies:
http://mathforum.org/k12/mathtips/mathtips.html

Because frequent calculator use tends to deprive students of a sense of numbers, they will often approach a sequence of multiplications and divisions the hard way. For instance, asked to calculate $770 \times 36/55$, they will first multiply 770 and 36 and then do a long division with the 55. They fail to recognize that both 770 and 55 can be divided by 11 and then by 5 to considerably simplify the problem. Give students plenty of practice in multiplying and dividing a sequence of integers and fractions so they are comfortable with canceling top and bottom terms.

Math language

There is an explosion of new math words as students enter the middle grades and start learning algebra and geometry.

This website provides an animated, colorfully illustrated dictionary of math terms:
http://www.amathsdictionaryforkids.com/

The following site is not colorful and animated but contains brief and clear definitions and many more advanced math terms:
http://www.amathsdictionaryforkids.com/

WEB LINKS

ALGEBRA
Algebra in bite-size pieces with quiz at the end
http://library.thinkquest.org/20991/alg/index.html
Algebra II: http://library.thinkquest.org/20991/alg2/index.html

Different levels plus quiz
http://www.math.com/homeworkhelp/Algebra.html

Clicking on the number leads to solution
http://www.math.armstrong.edu/MathTutorial/index.html

Algebraic Structures
Symbols and sets of numbers:
http://www.wtamu.edu/academic/anns/mps/math/mathlab/beg_algebra/beg_alg_t
ut2_sets.htm

Integers: http://amby.com/educate/math/integer.html
Card game to add and subtract integers: http://www.education-
world.com/a_tsl/archives/03-1/lesson001.shtml
Multiplying integers: http://www.aaastudy.com/mul65_x2.htm

Rational/irrational numbers: http://regentsprep.org/regents/math/math-
topic.cfm?TopicCode=rational

Several complex number exercise pages:
http://math.about.com/od/complexnumbers/Complex_Numbers.htm

Polynomial Equations and Inequalities
Systems of equations lessons and practice:
http://regentsprep.org/regents/math/math-topic.cfm?TopicCode=syslin
More practice:
http://www.sparknotes.com/math/algebra1/systemsofequations/problems3.rhtml
Word problems system of equations:
http://regentsprep.org/REgents/math/ALGEBRA/AE3/PracWord.htm
Inequalities: http://regentsprep.org/regents/Math/solvin/PSolvIn.htm
Inequality tutorial, examples, problems
http://www.wtamu.edu/academic/anns/mps/math/mathlab/beg_algebra/beg_alg_t
ut18_ineq.htm
Graphing linear inequalities tutorial
http://www.wtamu.edu/academic/anns/mps/math/mathlab/beg_algebra/beg_alg_t
ut24_ineq.htm
Quadratic equations tutorial, examples, problems
http://www.wtamu.edu/academic/anns/mps/math/mathlab/col_algebra/col_alg_tut
17_quad.htm

Practice factoring: http://regentsprep.org/Regents/math/math-topic.cfm?TopicCode=factor
Synthetic division tutorial:
http://www.wtamu.edu/academic/anns/mps/math/mathlab/col_algebra/col_alg_tut_37_syndiv.htm
Synthetic division Examples and problems: http://www.tpub.com/math1/10h.htm

Functions
Function, domain, range intro and practice
http://www.mathwarehouse.com/algebra/relation/math-function.php
Equations with rational expressions tutorial
http://www.wtamu.edu/academic/anns/mps/math/mathlab/col_algebra/col_alg_tut_15_rateq.htm
Practice with rational expressions
http://education.yahoo.com/homework_help/math_help/problem_list?id=minialg1gt_7_1
Practice simplifying radicals
http://www.bhs87.org/math/practice/radicals/radicalpractice.htm
Radical equations – lesson and practice
http://regentsprep.org/REgents/mathb/mathb-topic.cfm?TopicCode=7D3
Logarithmic functions tutorial
http://www.wtamu.edu/academic/anns/mps/math/mathlab/col_algebra/col_alg_tut_43_logfun.htm

Linear Algebra
Practice operations with matrices
http://www.castleton.edu/Math/finite/operation_practice.htm
Matrices, introduction and practice
http://www.math.csusb.edu/math110/src/matrices/basics.html
Vector practice tip: http://www.phy.mtu.edu/~suits/PH2100/vecdot.html

GEOMETRY
Geometry
http://library.thinkquest.org/20991/geo/index.html
http://www.math.com/students/homeworkhelp.html#geometry
http://regentsprep.org/Regents/math/geometry/math-GEOMETRY.htm

Parallelism
Parallel lines practice
http://www.algebralab.org/lessons/lesson.aspx?file=Geometry_AnglesParallelLinesTransversals.xml

Plane Euclidean Geometry
Geometry facts and practice http://www.aaaknow.com/geo.htm
Triangles intro and practice
http://www.staff.vu.edu.au/mcaonline/units/geometry/triangles.html
Polygons exterior and interior angles practice
http://regentsprep.org/Regents/Math/math-topic.cfm?TopicCode=poly
Angles in circles practice
http://regentsprep.org/Regents/math/geometry/GP15/PcirclesN2.htm
Congruence of triangles – lessons, practice
http://regentsprep.org/Regents/math/geometry/GP4/indexGP4.htm
Pythagorean theorem and converse
http://regentsprep.org/Regents/math/geometry/GP13/indexGP13.htm
Circle equation practice
http://www.regentsprep.org/Regents/math/algtrig/ATC1/circlepractice.htm
Interactive parabola http://www.mathwarehouse.com/geometry/parabola/
Ellipse practice problems http://www.mathwarehouse.com/ellipse/equation-of-ellipse.php#equationOfEllipse

Three-Dimensional Geometry
3D figures intro and examples
http://www.mathleague.com/help/geometry/3space.htm

Transformational Geometry
Interactive transformational geometry practice on coordinate plane
http://www.shodor.org/interactivate/activities/Transmographer/
Similar triangles practice
http://regentsprep.org/Regents/math/similar/PracSim.htm

http://www.algebralab.org/practice/practice.aspx?file=Geometry_UsingSimilarTriangles.xml

NUMBER THEORY
Natural Numbers
http://online.math.uh.edu/MiddleSchool/Vocabulary/NumberTheoryVocab.pdf
GCF and LCM practice
http://teachers.henrico.k12.va.us/math/ms/C1Files/01NumberSense/1_5/6035prac.htm

PROBABILITY AND STATISTICS
Probability
Probability intro and practice
http://www.mathgoodies.com/lessons/vol6/intro_probability.html
Permutation and combination practice
http://www.regentsprep.org/Regents/math/algtrig/ATS5/PCPrac.htm
Conditional probability problems
http://homepages.ius.edu/MEHRINGE/T102/Supplements/HandoutConditionalProbability.htm

Statistics
Statistics lessons and interactive practice
http://www.aaaknow.com/sta.htm
Range, mean, median, mode exercises
http://www.mathgoodies.com/lessons/vol8/practice_vol8.html
http://regentsprep.org/regents/Math/mean/Pmeasure.htm

Sample Test 1

1) Given W = whole numbers
N = natural numbers
Z = integers
R = rational numbers
I = irrational numbers
Which of the following is not true?
(Skill 1.1) (Easy Rigor)

A) $R \subset I$

B) $W \subset Z$

C) $Z \subset R$

D) $N \subset W$

2) Given that *x*, *y*, and *z* are prime numbers, which of the following is true?
(Skill 1.2)(Easy Rigor)

A) $x + y$ is always prime

B) xyz is always prime

C) xy is sometimes prime

D) $x + y$ is sometimes prime

3) Find the GCF of $2^2 \cdot 3^2 \cdot 5$ and $2^2 \cdot 3 \cdot 7$.
(Skill 1.4) (Average Rigor)

A) $2^5 \cdot 3^3 \cdot 5 \cdot 7$

B) $2 \cdot 3 \cdot 5 \cdot 7$

C) $2^2 \cdot 3$

D) $2^3 \cdot 3^2 \cdot 5 \cdot 7$

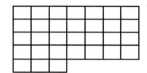

4) The above diagram would be most appropriate for illustrating which of the following?
(Skill 2.1) (Average Rigor)

A) $7 \times 4 + 3$

B) $31 \div 8$

C) 28×3

D) $31 - 3$

5) 2^{-3} is equivalent to
(Skill 2.2)(Average Rigor)

A)   .8

B)   -.8

C)   125

D)   0.125

6) Simplify: $\sqrt{27} + \sqrt{75}$
(Skill 2.3)(Average Rigor)

A) $8\sqrt{3}$

B)   34

C) $34\sqrt{3}$

D) $15\sqrt{3}$

7) The mass of a cookie is closest to
(Skill 2.4)(Average Rigor)

A) 0.5 kg

B) 0.5 grams

C) 15 grams

D) 1.5 grams

8) A sofa sells for $520. If the retailer makes a 30% profit, what was the wholesale price?
(Skill 2.5)(Rigorous)

A) $400

B) $676

C) $490

D) $364

9) Which of the following does not correctly relate an inverse operation?
(Skill 3.1)(Average Rigor)

A) $a - b = a + -b$

B) $a \times b = b \times -a$

C) $\sqrt{a^2} = a^2$

D) $a \times \dfrac{1}{a} = 1$

10) Which axiom is incorrectly applied?
(Skill 3.2)(Average Rigor)

$3x + 4 = 7$

Step a. $3x + 4 - 4 = 7 - 4$
additive equality

Step b. $3x + 4 - 4 = 3$
commutative axiom of addition

Step c. $3x + 0 = 3$
additive inverse

Step d. $3x = 3$
additive identity

A) step a

B) step b

C) step c

D) step d

11) Evaluate $3^x \times 3^{2y}$ where x is 1 and y is 2.
(Skill 3.3)(Average Rigor)

A) 243

B) 81

C) 729

D) 9

12) Evaluate $3^{1/2}(9^{1/3})$
(Skill 3.3)(Rigorous)

A) $27^{5/6}$

B) $9^{7/12}$

C) $3^{5/6}$

D) $3^{6/7}$

13) Simplify: $\dfrac{3.5\times10^{-10}}{0.7\times10^{4}}$
(Skill 3.3) (Rigorous)

A) 0.5×10^{6}

B) 5.0×10^{-6}

C) 5.0×10^{-14}

D) 0.5×10^{-14}

14) You have a gallon of water and remove a total of 30 ounces. How many milliliters do you have left?
(Skill 4.2) (Rigorous)

A) 2,900 mL

B) 1,100 mL

C) 980 mL

D) 1,000 mL

15) If a circle has an area of 25 cm^2, what is its circumference to the nearest tenth of a centimeter?
(Skill 4.4)(Average Rigor)

A) 78.5 cm

B) 17.7 cm

C) 8.9 cm

D) 15.7 cm

16) Find the area of the figure below.
(Skill 4.4)(Rigorous)

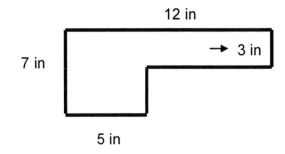

A) 56 in.^{2}

B) 27 in.^{2}

C) 71 in.^{2}

D) 170 in.^{2}

17) Determine the area of the shaded region of the trapezoid in terms of *x* and *y* *(the height of* $\triangle ABC$). $\overline{DE} =$ **2x &** $\overline{DC} = $ **3x.**
(Skill 4.4)(Rigorous)

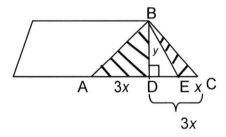

A) $4xy$

B) $2xy$

C) $3x^2y$

D) There is not enough information given.

18) What happens to the volume of a square pyramid when the sides of the base are tripled?
(Skill 4.6) (Rigorous)

A) The volume is increased 9 times.

B) The volume is increased 8 times.

C) The volume is increased 27 times.

D) The volume is increased 16 times.

19) Two non-coplanar lines that do not intersect are labeled
(Skill 5.1)(Easy Rigor)

A) parallel lines

B) perpendicular lines

C) skew lines

D) alternate exterior lines

20) Choose the diagram that illustrates the construction of a perpendicular line to the line at a given point.
(Skill 5.2) (Average Rigor)

A)

B)

C)

D)

21) Which postulate or theorem can be used to prove $\triangle BAK \cong \triangle MKA$?
(Skill 5.3) (Rigorous)

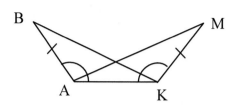

A)   SSS

B)   ASA

C)   SAS

D)   AAS

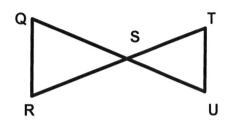

22) Given $QS \cong TS$ and $RS \cong US$, prove $\triangle QRS \cong \triangle TUS$.

1) $QS \cong TS$ 1) Given
2) $RS \cong US$ 2) Given
3) $\angle TSU \cong \angle QSR$ 3) ?
4) $\triangle TSU \cong \triangle QSR$ 4) SAS

Give the reason that justifies step 3.
(Skill 5.3)(Rigorous)

A) Congruent parts of congruent triangles are congruent

B) Reflexive axiom of equality

C) Alternate interior angle Theorem

D) Vertical angle theorem

23) Given that $QO \perp NP$ and $QO \cong NP$, quadrilateral *NOPQ* can most accurately be described as a
(Skill 5.4)(Average Rigor)

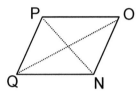

A) parallelogram

B) rectangle

C) square

D) rhombus

24) If a ship sails due south 6 miles, then due west 8 miles, how far was it from the starting point?
(Skill 5.5)(Rigorous)

A) 100 miles

B) 10 miles

C) 14 miles

D) 48 miles

25) Find the center of a circle with a diameter whose endpoints are (4, 5) and (-4, -6).
(Skill 6.2)(Average Rigor)

A) (-2, ½)

B) (0, -1/2)

C) (-1, 0)

D) (0, 1)

26) Find the distance between points (3, 7) and (-3, 4).
(Skill 6.2)(Rigorous)

A)   9

B)   45

C) $3\sqrt{5}$

D) $5\sqrt{3}$

27) Given similar polygons with corresponding sides of lengths 9 units and 15 units, find the perimeter of the smaller polygon if the perimeter of the larger polygon is 150 units.
(Skill 6.4)(Rigorous)

A)   54 units

B)   135 units

C)   90 units

D)   126 units

28) {1, 4, 7, 10, . . .}.
What is the 40th term in this sequence?
(Skill 7.1)(Rigorous)

A)   43

B)   121

C)   118

D)   120

29) Which set illustrates a function?
(Skill 7.3)(Average Rigor)

A)   {(0,1) (0,2) (0,3) (0,4)}

B)   {(3,9) (-3,9) (4,16) (-4,16)}

C)   {(1,2) (2,3) (3,4) (1,4)}

D)   {(2,4) (3,6) (4,8) (4,16)}

30) The volume of water flowing through a pipe varies directly with the square of the radius of the pipe. If the water flows at a rate of 80 liters per minute through a pipe with a radius of 4 cm, at what rate would water flow through a pipe with a radius of 3 cm?
(Skill 7.4)(Rigorous)

A)   45 liters per minute

B)   6.67 liters per minute

C)   60 liters per minute

D)   4.5 liters per minute

31) The discriminant of a quadratic equation is evaluated and determined to be -3. The equation has
(Skill 7.5) (Easy Rigor)

A) one real root

B) one complex root

C) two roots, both real

D) two roots, both complex

32) Which of the following is incorrect?
(Skill 8.1)(Average Rigor)

A) $(x^2y^3)^2 = x^4y^6$

B) $m^2(2n)^3 = 8m^2n^3$

C) $(m^3n^4)/(m^2n^2) = mn^2$

D) $(x+y^2)^2 = x^2 + y^4$

33) Which of the following is a factor of $k^3 - m^3$?
(Skill 8.1)(Rigorous)

A) $k^2 + m^2$

B) $k + m$

C) $k^2 - m^2$

D) $k - m$

34) Solve for x: $18 = 4 + |2x|$
(Skill 8.3)(Easy Rigor)

A) $\{-11, 7\}$

B) $\{-7, 0, 7\}$

C) $\{-7, 7\}$

D) $\{-11, 11\}$

35) Three less than four times a number is five times the sum of that number and 6. Which equation could be used to solve this problem?
(Skill 8.4)(Average Rigor)

A) $3 - 4n = 5(n + 6)$

B) $3 - 4n + 5n = 6$

C) $4n - 3 = 5n + 6$

D) $4n - 3 = 5(n + 6)$

36) If three cups of concentrate are needed to make 2 gallons of fruit punch, how many cups are needed to make 5 gallons?
(Skill 8.5)(Average Rigor)

A) 6 cups

B) 7 cups

C) 7.5 cups

D) 10 cups

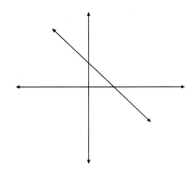

37) Which equation is represented by the above graph?
(Skill 9.2)(Average Rigor)

A) $x - y = 3$

B) $x - y = -3$

C) $x + y = 3$

D) $x + y = -3$

38) What is the solution to the system of equations?

$3x + 2y = 12$
$12x + 8y = 15$
(Skill 9.5)(Rigorous)

A)   all real numbers

B) $x = 4, y = 4$

C) $x = 2, y = -1$

D)   No solution

39) A boat travels 30 miles upstream in three hours. It makes the return trip in one and a half hours. What is the speed of the boat in still water?
(Skill 9.5)(Rigorous)

A)   10 mph

B)   15 mph

C)   20 mph

D)   30 mph

40) What is the slope of the equation $2y = 4x + 3$?
(Skill 9.6)(Average Rigor)

A)   2

B)   3

C)   1/2

D)   4

41) Graph the solution:
(Skill 9.7)(Average Rigor)

$|x| + 7 < 13$

A)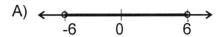

B)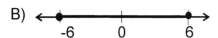

C)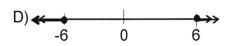

D)

42) The following is not an important consideration in statistical sampling:
(Rigorous)(Skill 10.1)

A) Sample size

B) Method of sample selection

C) Type of question asked

D) Number of investigators

43) Which type of graph uses symbols to represent quantities?
(Skill 10.2) (Easy)

A) Bar graph

B) Line graph

C) Pictograph

D) Circle graph

44)	Which statement is true about George's budget?
(Skill 10.3)(Easy Rigor)

A)	George spends the greatest portion of his income on food.

B)	George spends twice as much on utilities as he does on his mortgage.

C)	George spends twice as much on utilities as he does on food.

D)	George spends the same amount on food and utilities as he does on mortgage.

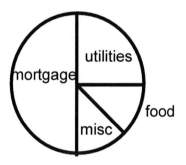

45)	Find the median of the following set of data:
14 3 7 6 11 20
(Skill 10.4) (Easy Rigor)

A)	9

B)	8.5

C)	7

D)	11

46) Half the students in a class scored 80% on an exam, most of the rest scored 85% except for one student who scored 10%. Which would be the best measure of central tendency for the test scores?
(Skill 10.4)(Easy Rigor)

A) mean

B) median

C) mode

D) either the median or the mode because they are equal

47) If the correlation between two variables is given as zero, the association between the two variables is
(Skill 10.6) (Rigorous)

A) negative linear

B) positive linear

C) quadratic

D) random

48) Which of the following sets is closed under division?
(Skill 11.1)(Average Rigor)

I)   {½, 1, 2, 4}

II)   {-1, 1}

III) {-1, 0, 1}

A) I only

B) II only

C) III only

D) I and II

49) How many ways are there to choose a potato and two green vegetables from a choice of three potatoes and seven green vegetables?
(Skill 11.2)(Rigorous)

A) 126

B) 63

C) 21

D) 252

50) If a horse will probably win three races out of ten, what are the odds that he will win?
(Skill 11.3)(Average Rigor)

A) 3:10

B) 7:10

C) 3:7

D) 7:3

51) Given a drawer with 5 black socks, 3 blue socks, and 2 red socks, what is the probability that you will draw two black socks in two draws in a dark room?
(Skill 11.3)(Rigorous)

A)   2/9

B)   1/4

C)   17/18

D)   1/18

52) Choose the least appropriate set of manipulatives for a sixth grade class.
(Skill 12.1) (Easy)

A)   graphing calculators, compasses, rulers, conic section models

B)   two color counters, origami paper, markers, yarn, balance, meter stick, colored pencils, beads

C)   balance, meter stick, colored pencils, beads

D)   paper cups, beans, tangrams, geoboards

53) A teacher plans an activity that involves students calculating how many chair legs are in the classroom, given that there are 30 chairs and each chair has 4 legs. This activity is introducing the ideas of:
(Skill 12.1)(Average rigor)

A) Probability

B) Statistics

C) Geometry

D) Algebra

54) If 80 students are in sports and 100 students are in band with 20 students in both band and sports, identify the appropriate pictorial represnation.
(Skill 12.2) (Average Rigor)

A)
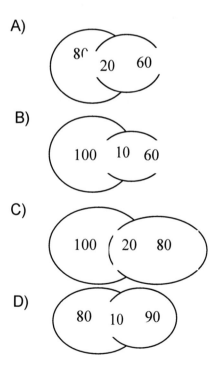

B)

C)

D)

55) Tom's age is one less than half the sum of his brother's and sister's ages. Tom's brother is twice as old as Tom, and his sister is 2 years younger than Tom. This information can be expressed symbolically in the following way:
(Skill 12.3) (Rigorous)

A) $n = \frac{1}{2}(2n + n - 2) - 1$

B) $n = \frac{1}{2}(2n + n - 2) + 1$

C) $n = \frac{1}{2}(2n + n - 2 - 1)$

D) $n = \frac{1}{2}(2n + n - 2 + 1)$

56) What is the ratio of the frequencies of the third and sixth notes in a musical octave given that the frequency of a note = frequency of starting note $\times 2^{N/12}$ where N is the number of notes from the starting point.
(Skill 12.5) (Rigorous)

A) 1.19

B) 1.41

C) 0.71

D) 0.84

57) When you begin by assuming the conclusion of a theorem is false, then show that through a sequence of logically correct steps you contradict an accepted fact, this is known as
(Easy)(Skill 13.1)

A) inductive reasoning

B) direct proof

C) indirect proof

D) exhaustive proof

58) Which of the following statements is logically equivalent to
"It is not true that Gina owns no books"
(Average Rigor)(Skill 13.3)

A) It is true that Gina owns a lot of books

B) It is likely that Gina owns some books

C) It is true that Gina owns books

D) None of the above

59) **Which of the following arguments is valid? (Rigorous)(Skill 13.3)**

A) People who support tax reduction will vote for Mr.Allen. Mr.Brown voted for Mr.Allen. Conclusion: Mr. Brown supports tax reduction.

B) If Mr.Allen is elected, he will reduce taxes. Mr. Allen was elected. Conclusion: Taxes were reduced.

C) If Mr.Allen is elected, he will reduce taxes. Taxes were reduced. Conclusion: Mr. Allen was elected.

D) None of the above

60) **Which is the least appropriate strategy to emphasize when teaching problem solving? (Easy)(Skill 13.4)**

A) guess and check

B) look for key words to indicate operations such as all together—add, more than, subtract, times, multiply

C) make a diagram

D) solve a simpler version of the problem

Answer Key

1. A	17. B	33. D	49. A
2. D	18. A	34. C	50. C
3. C	19. C	35. D	51. A
4. A	20. D	36. C	52. A
5. D	21. C	37. C	53. D
6. A	22. D	38. D	54. A
7. C	23. C	39. B	55. A
8. A	24. B	40. A	56. D
9. B	25. B	41. A	57. C
10. B	26. C	42. D	58. C
11. A	27. C	43. C	59. B
12. B	28. C	44. C	60. B
13. C	29. B	45. A	
14. A	30. A	46. B	
15. B	31. D	47. D	
16. A	32. D	48. B	

Rigor Table

Easy Rigor 20%	Average Rigor 40%	Rigorous 40%
1, 2, 19, 31, 34, 43, 44, 45, 46, 52, 57, 60	3, 4, 5, 6, 7, 9, 10, 11, 15, 20, 23, 25, 29, 32, 35, 36, 37, 40, 41, 48, 50, 53, 54, 58	8, 12, 13, 14, 16, 17, 18, 21, 22, 24, 26, 27, 28, 30, 33, 38, 39, 42, 47, 49, 51, 55, 56, 59

Rationales for Sample Test 1

1) Given W = whole numbers
   N = natural numbers
   Z = integers
   R = rational numbers
   I = irrational numbers

   Which of the following is not true?
   (Skill 1.1) (Easy Rigor)

   A) $R \subset I$

   B) $W \subset Z$

   C) $Z \subset R$

   D) $N \subset W$

   Answer: A

   The rational numbers are not a subset of the irrational numbers. All of the other statements are true.

2) **Given that x, y, and z are prime numbers, which of the following is true?**
   (Skill 1.2)(Easy Rigor)

   A) $x + y$ is always prime

   B) xyz is always prime

   C) xy is sometimes prime

   D) $x + y$ is sometimes prime

   Answer: D

   $x + y$ is sometimes prime. B and C show the products of two numbers which are always composite. $x + y$ may sometimes be prime but not always (e.g. $3 + 2 = 5$ is prime but $3 + 3 = 6$ is not).

3)   Find the GCF of $2^2 \cdot 3^2 \cdot 5$ and $2^2 \cdot 3 \cdot 7$.
     (Skill 1.4) (Average Rigor)

   A) $2^5 \cdot 3^3 \cdot 5 \cdot 7$

   B) $2 \cdot 3 \cdot 5 \cdot 7$

   C) $2^2 \cdot 3$

   D) $2^3 \cdot 3^2 \cdot 5 \cdot 7$

   Answer: C

   Choose the highest common exponent for each prime factor.

4)   **The above diagram would be most appropriate for illustrating which of the following?**
     (Skill 2.1) (Average Rigor)

   A) $7 \times 4 + 3$

   B) $31 \div 8$

   C) 28×3

   D) $31 - 3$

   Answer: A

   A shows a 7x4 rectangle with 3 additional units. C is inappropriate. B is the division based on 8 which makes no sense. D could show how subtraction might be visualized leaving a composite difference but there is nothing indicating that the last 3 units are being subtracted.

5) **2^{-3} is equivalent to**
    (Skill 2.2)(Average Rigor)

    A)   .8

    B)   -.8

    C)   125

    D)   0.125

    Answer: D

    Express 2^{-3} as the fraction 1/8, then convert to a decimal.

6) **Simplify: $\sqrt{27} + \sqrt{75}$**
    (Skill 2.3)(Average Rigor)

    A) $8\sqrt{3}$

    B)   34

    C) $34\sqrt{3}$

    D) $15\sqrt{3}$

    Answer: A

    Simplifying radicals gives $\sqrt{27} + \sqrt{75} = 3\sqrt{3} + 5\sqrt{3} = 8\sqrt{3}$.

7) **The mass of a cookie is closest to**
    (Skill 2.4)(Average Rigor)

    A)   0.5 kg

    B)   0.5 grams

    C)   15 grams

    D)   1.5 grams

    Answer: C

    In terms of commonly used U.S. units, 15 grams is about half an ounce and 0.5 kg is about a pound.

8) **A sofa sells for $520. If the retailer makes a 30% profit, what was the wholesale price?**
(Skill 2.5)(Rigorous)

A) $400

B) $676

C) $490

D) $364

Answer: A

Let x be the wholesale price, then x + .30x = 520, 1.30x = 520. Divide both sides by 1.30.

9) **Which of the following does not correctly relate an inverse operation?**
(Skill 3.1)(Average Rigor)

A) $a - b = a + -b$

B) $a \times b = b \times -a$

C) $\sqrt{a^2} = a^2$

D) $a \times \dfrac{1}{a} = 1$

Answer: B

B is incorrect. A, C, and D illustrate various properties of inverse relations.

10)   **Which axiom is incorrectly applied?**
      (Skill 3.2)(Average Rigor)

$3x + 4 = 7$

Step a. $3x + 4 - 4 = 7 - 4$
additive equality

Step b. $3x + 4 - 4 = \ 3$
commutative axiom of addition

Step c. $3x + 0 = \ 3$
additive inverse

Step d. $3x \ = \ 3$
additive identity

A)   step a

B)   step b

C)   step c

D)   step d

Answer: B

In simplifying from step a to step b, 3 replaced 7 – 4, therefore the correct justification would be subtraction or substitution.

11)   **Evaluate 3^x x 3^{2y} where x is 1 and y is 2.**
      (Skill 3.3)(Average Rigor)

A)   243

B)   81

C)   729

D)   9

Answer: A

Since a^x x $a^y = a^{(x + y)}$, 3^x x $3^{2y} = 3^{(1 + 4)} = 3^5 = 243$.

12) **Evaluate** $3^{1/2}(9^{1/3})$
   (Skill 3.3)(Rigorous)

   A) $27^{5/6}$

   B) $9^{7/12}$

   C) $3^{5/6}$

   D) $3^{6/7}$

   Answer: B

   Getting the bases the same gives us $3^{\frac{1}{2}}3^{\frac{2}{3}}$. Adding exponents gives $3^{\frac{7}{6}}$. Then some additional manipulation of exponents produces
   $3^{\frac{7}{6}} = 3^{\frac{14}{12}} = \left(3^2\right)^{\frac{7}{12}} = 9^{\frac{7}{12}}$.

13) **Simplify:** $\dfrac{3.5\times10^{-10}}{0.7\times10^{4}}$
   (Skill 3.3) (Rigorous)

   A) 0.5×10^{6}

   B) 5.0×10^{-6}

   C) 5.0×10^{-14}

   D) 0.5×10^{-14}

   Answer: C

   Divide the decimals and subtract the exponents.

14) **You have a gallon of water and remove a total of 30 ounces. How many milliliters do you have left?**
(Skill 4.2) (Rigorous)

A)   2,900 mL

B)   1,100 mL

C)   980 mL

D)   1,000 mL

Answer: A

1 gallon = 128 fluid ouces. If 30 ounces are removed, you have 98 ounces left. Since 1 fluid ounce = 29.6 mL, 98 ounces = 2,900 mL.

15) **If a circle has an area of 25 cm^2, what is its circumference to the nearest tenth of a centimeter?**
(Skill 4.4)(Average Rigor)

A)   78.5 cm

B)   17.7 cm

C)   8.9 cm

D)   15.7 cm

Answer: B

Find the radius by solving $\Pi r^2 = 25$. Then substitute $r = 2.82$ into $C = 2\Pi r$ to obtain the circumference.

16) **Find the area of the figure below.**
   (Skill 4.4)(Rigorous)

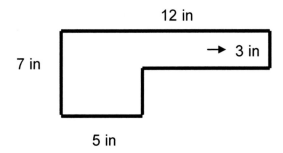

A)   56 in.^{2}

B)   27 in.^{2}

C)   71 in.^{2}

D)   170 in.^{2}

Answer: A

Divide the figure into two rectangles, 7 in. x 5 in. and 7 in. x 3 in. The combined area = 35 in.^{2} + 21 in.^{2} = 56 in.^{2}

17) Determine the area of the shaded region of the trapezoid in terms of *x* and *y* (the height of $\triangle ABC$). $\overline{DE}$ = **2x** & $\overline{DC}$ = **3x**.
(Skill 4.4)(Rigorous)

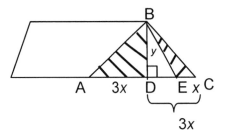

A)　　$4xy$

B)　　$2xy$

C)　　$3x^2y$

D)　　There is not enough information given.

Answer: B

To find the area of the shaded region, find the area of triangle *ABC* and then subtract the area of triangle *DBE*. The area of triangle ABC is .5(6x)(y) = 3xy. The area of triangle *DBE* is .5(2x)(y) = xy. The difference is 2xy.

18) What happens to the volume of a square pyramid when the sides of the base are tripled?
(Skill 4.6) (Rigorous)

A)　　The volume is increased 9 times.

B)　　The volume is increased 8 times.

C)　　The volume is increased 27 times.

D)　　The volume is increased 16 times.

Answer: A

The area of the base is increased 9 times, which increases the volume 9 times when the sides are tripled.

19) Two non-coplanar lines that do not intersect are labeled
(Skill 5.1)(Easy Rigor)

    A)   parallel lines

    B)   perpendicular lines

    C)   skew lines

    D)   alternate exterior lines

Answer: C

20) **Choose the diagram that illustrates the construction of a perpendicular line to the line at a given point.**
(Skill 5.2) (Average Rigor)

A)

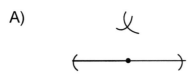

B)

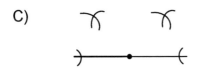

C)

D)

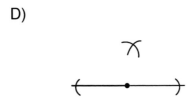

Answer: D

Given a point on a line, place the compass point there and draw two arcs intersecting the line in two points, one on either side of the given point. Then using any radius larger than half the new segment produced, and with the pointer at each end of the new segment, draw arcs that intersect above the line. Connect this new point with the given point.

21)	Which postulate or theorem can be used to prove $\triangle BAK \cong \triangle MKA$? (Skill 5.3) (Rigorous)

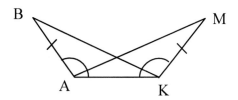

A)	SSS

B)	ASA

C)	SAS

D)	AAS

**Answer:	C**

Since side *AK* is common to both triangles, the triangles can be proved congruent by using the Side-Angle-Side Postulate.

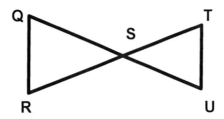

22) Given **QS ≅ TS** and **RS ≅ US**, prove △**QRS ≅** △**TUS**.

l) $QS \cong TS$ 1) Given
2) $RS \cong US$ 2) Given
3) $\angle TSU \cong \angle QSR$ 3) ?
4) $\triangle TSU \cong \triangle QSR$ 4) SAS

Give the reason that justifies step 3.

(Skill 5.3)(Rigorous)

A) Congruent parts of congruent triangles are congruent

B) Reflexive axiom of equality

C) Alternate interior angle Theorem

D) Vertical angle theorem

Answer: D

Non-adjacent angles formed by intersecting lines are called vertical angles and are congruent.

23) Given that $QO \perp NP$ and $QO \cong NP$, quadrilateral *NOPQ* can most accurately be described as a
(Skill 5.4)(Average Rigor)

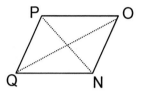

A) parallelogram

B) rectangle

C) square

D) rhombus

Answer: C

In an ordinary parallelogram, the diagonals are not perpendicular or equal in length. In a rectangle, the diagonals are not necessarily perpendicular. In a rhombus, the diagonals are not equal in length. In a square, the diagonals are both perpendicular and congruent.

24) If a ship sails due south 6 miles, then due west 8 miles, how far was it from the starting point?
(Skill 5.5)(Rigorous)

A) 100 miles

B) 10 miles

C) 14 miles

D) 48 miles

Answer: B

Draw a right triangle with legs of 6 and 8. Find the hypotenuse using the Pythagorean Theorem. $6^2 + 8^2 = c^2$. Therefore, $c = 10$ miles.

25) **Find the center of a circle with a diameter whose endpoints are (4, 5) and (-4, -6).**
(Skill 6.2)(Average Rigor)

A)   (-2, ½)

B)   (0, -1/2)

C)   (-1, 0)

D)   (0, 1)

Answer: B

The center of the circle is the midpoint of the diameter and can be found using the midpoint formula.

$$\frac{(4 + (-4)),}{2} \quad \frac{5 + (-6))}{2} = (0, -1/2)$$

26) **Find the distance between points (3, 7) and (-3, 4).**
(Skill 6.2)(Rigorous)

A)   9

B)   45

C) $3\sqrt{5}$

D) $5\sqrt{3}$

Answer: C

Using the distance formula
$$\sqrt{[(-3) - 3]^2 + (4 - 7)^2}$$
$$= \sqrt{36 + 9} = \sqrt{45} = 3\sqrt{5}$$

27) Given similar polygons with corresponding sides of lengths 9 units and 15 units, find the perimeter of the smaller polygon if the perimeter of the larger polygon is 150 units.
(Skill 6.4)(Rigorous)

   A) 54 units

   B) 135 units

   C) 90 units

   D) 126 units

   Answer: C

   The perimeters of similar polygons are directly proportional to the lengths of their sides, therefore $9/15 = x/150$. Cross multiply to obtain $1,350 = 15x$, then divide by 15 to obtain the perimeter of the smaller polygon.

28) {1, 4, 7, 10, . . .}.
What is the 40th term in this sequence?
(Skill 7.1)(Rigorous)

   A) 43

   B) 121

   C) 118

   D) 120

   Answer: C

   This is an arithmetic sequence with first term 1 and common difference 3.

   Using the formula $a_n = a_1 + (n-1)d$, the 40th term $= 1 + (40-1)3 = 118$.

29) **Which set illustrates a function?**
    (Skill 7.3)(Average Rigor)

    A)   {(0,1) (0,2) (0,3) (0,4)}

    B)   {(3,9) (-3,9) (4,16) (-4,16)}

    C)   {(1,2) (2,3) (3,4) (1,4)}

    D)   {(2,4) (3,6) (4,8) (4,16)}

    Answer: B

    Each number in the domain can only be matched with one number in the range. A is not a function because 0 is mapped to 4 different numbers in the range. In C, 1 is mapped to two different numbers. In D, 4 is also mapped to two different numbers.

30) **The volume of water flowing through a pipe varies directly with the square of the radius of the pipe. If the water flows at a rate of 80 liters per minute through a pipe with a radius of 4 cm, at what rate would water flow through a pipe with a radius of 3 cm?**
    (Skill 7.4)(Rigorous)

    A)   45 liters per minute

    B)   6.67 liters per minute

    C)   60 liters per minute

    D)   4.5 liters per minute

    Answer: A

    Set up the direct variation: $\dfrac{V_1}{r_1^2} = \dfrac{V_2}{r_2^2}$. Substituting gives $\dfrac{80}{16} = \dfrac{V_2}{9}$.

    Solving for V_2 gives 45 liters per minute.

31) **The discriminant of a quadratic equation is evaluated and determined to be -3. The equation has**
    (Skill 7.5) (Easy Rigor)

    A)   one real root

    B)   one complex root

    C)   two roots, both real

    D)   two roots, both complex

Answer: D

The discriminant is the number under the radical sign in the quadratic formula. Since it is negative, the two roots of the equation are complex.

32) **Which of the following is incorrect?**
    (Skill 8.1)(Average Rigor)

    A) $(x^2 y^3)^2 = x^4 y^6$

    B) $m^2 (2n)^3 = 8m^2 n^3$

    C) $(m^3 n^4)/(m^2 n^2) = mn^2$

    D) $(x + y^2)^2 = x^2 + y^4$

Answer: D

Using FOIL to do the expansion, we get $(x + y^2)^2 = (x + y^2)(x + y^2) = x^2 + 2xy^2 + y^4$.

33) **Which of the following is a factor of** $k^3 - m^3$ **?**
   (Skill 8.1)(Rigorous)

   A) $k^2 + m^2$

   B) $k + m$

   C) $k^2 - m^2$

   D) $k - m$

   Answer: D

   The complete factorization for a difference of cubes is $(k - m)(k^2 + mk + m^2)$.

34) **Solve for** x: $18 = 4 + |2x|$
   (Skill 8.3)(Easy Rigor)

   A) $\{-11, 7\}$

   B) $\{-7, 0, 7\}$

   C) $\{-7, 7\}$

   D) $\{-11, 11\}$

   Answer: C

   Using the definition of absolute value, two equations are possible: $18 = 4 + 2x$ and $18 = 4 - 2x$. Solving for x gives $x = 7$ and $x = -7$.

35) **Three less than four times a number is five times the sum of that number and 6. Which equation could be used to solve this problem? (Skill 8.4)(Average Rigor)**

    A) $3 - 4n = 5(n + 6)$

    B) $3 - 4n + 5n = 6$

    C) $4n - 3 = 5n + 6$

    D) $4n - 3 = 5(n + 6)$

Answer: D

Be sure to enclose the sum of the number and 6 in parentheses.

36) **If three cups of concentrate are needed to make 2 gallons of fruit punch, how many cups are needed to make 5 gallons? (Skill 8.5)(Average Rigor)**

    A)   6 cups

    B)   7 cups

    C)   7.5 cups

    D)   10 cups

Answer: C

Set up the proportion $3/2 = x/5$, cross multiply to obtain $15 = 2x$, and solve for x. to get $x = 15/2 = 7.5$.

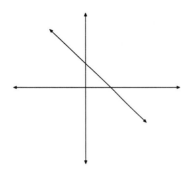

37) **Which equation is represented by the above graph?**
(Skill 9.2)(Average Rigor)

A) $x - y = 3$

B) $x - y = -3$

C) $x + y = 3$

D) $x + y = -3$

Answer: C

By looking at the graph, we can determine the slope to be negative and the *y*-intercept to be positive. The equation C is the only one that meets both criteria.

38) **What is the solution to the system of equations?**

$3x + 2y = 12$
$12x + 8y = 15$
(Skill 9.5)(Rigorous)

A)   all real numbers

B) $x = 4, y = 4$

C) $x = 2, y = -1$

D)   No solution

Answer: D

Multiplying the top equation by -4 and adding results in the equation $0 = -33$. Since this is a false statement, there is no solution. Furthermore, the equations of both lines have the same slope, so they are parallel and never intersect.

39) A boat travels 30 miles upstream in three hours. It makes the return trip in one and a half hours. What is the speed of the boat in still water?
(Skill 9.5)(Rigorous)

A) 10 mph

B) 15 mph

C) 20 mph

D) 30 mph

Answer: B

Let x = the speed of the boat in still water and c = the speed of the current.

	rate	time	distance
upstream	$x - c$	3	30
downstream	$x + c$	1.5	30

Solve the system:
$$3x - 3c = 30$$
$$1.5x + 1.5c = 30$$

Multiply the 2nd equation by 2, add the two equations and solve for x.

40) What is the slope of the equation $2y = 4x + 3$?
(Skill 9.6)(Average Rigor)

A) 2

B) 3

C) 1/2

D) 4

Answer: A

$y = 2x + 3/2$, so the slope = 2

41) Graph the solution:
(Skill 9.7)(Average Rigor)

$|x| + 7 < 13$

A)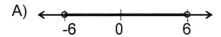

B)

C)

D)

Answer: A

Solve by adding -7 to each side of the inequality. Since the absolute value of x is less than 6, x must be between -6 and 6. The end points are not included so the circles on the graph are hollow.

42) The following is not an important consideration in statistical sampling:
(Rigorous)(Skill 10.1)

A) Sample size

B) Method of sample selection

C) Type of question asked

D) Number of investigators

Answer: D

The results of statistical sampling can vary greatly depending on the size of the sample, how the sample is selected and what types of questions are asked. Therefore, A, B and C are all important considerations in statistical sampling.

43) Which type of graph uses symbols to represent quantities? (Skill 10.2) (Easy)

    A)   Bar graph

    B)   Line graph

    C)   Pictograph

    D)   Circle graph

Answer: C

A pictograph shows comparison of quantities using symbols. Each symbol represents a number of items.

44) Which statement is true about George's budget? (Skill 10.3)(Easy Rigor)

    A)   George spends the greatest portion of his income on food.

    B)   George spends twice as much on utilities as he does on his mortgage.

    C)   George spends twice as much on utilities as he does on food.

    D)   George spends the same amount on food and utilities as he does on mortgage.

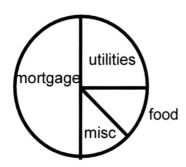

Answer: C

George spends twice as much on utilities than on food.

45) **Find the median of the following set of data:**
14 3 7 6 11 20
(Skill 10.4) (Easy Rigor)

A) 9

B) 8.5

C) 7

D) 11

Answer: A

Place the numbers is ascending order: 3 6 7 11 14 20. Find the average of the two middle numbers (7 + 11)1/2 = 9

46) **Half the students in a class scored 80% on an exam, most of the rest scored 85% except for one student who scored 10%. Which would be the best measure of central tendency for the test scores?**
(Skill 10.4)(Easy Rigor)

A) mean

B) median

C) mode

D) either the median or the mode because they are equal

Answer: B

In this set of data, the median would be the most representative measure of central tendency since the median is independent of extreme values. Because of the 10% outlier, the mean (average) would be disproportionately skewed. In this data set, it is true that the median and the mode (number which occurs most often) are the same, but the median remains the best choice because of its special properties.

47) **If the correlation between two variables is given as zero, the association between the two variables is**
(Skill 10.6) (Rigorous)

A) negative linear

B) positive linear

C) quadratic

D) random

Answer: D

A correlation of 1 indicates a perfect positive linear association, a correlation of -1 indicates a perfect negative linear association while a correlation of zero indicates a random relationship between the variables.

48) **Which of the following sets is closed under division?**
(Skill 11.1)(Average Rigor)

I) $\{½, 1, 2, 4\}$

II) $\{-1, 1\}$

III) $\{-1, 0, 1\}$

A) I only

B) II only

C) III only

D) I and II

Answer: B

I is not closed because $\dfrac{4}{.5} = 8$ and 8 is not in the set.

III is not closed because $\dfrac{1}{0}$ is undefined.

II is closed because $\dfrac{-1}{1} = -1, \dfrac{1}{-1} = -1, \dfrac{1}{1} = 1, \dfrac{-1}{-1} = 1$ and all the answers are in the set.

49) **How many ways are there to choose a potato and two green vegetables from a choice of three potatoes and seven green vegetables?**
(Skill 11.2)(Rigorous)

A)   126

B)   63

C)   21

D)   252

Answer: A

There are 3 slots to fill. There are 3 choices for the first, 7 for the second, and 6 for the third. Therefore, the total number of choices is 3(7)(6) = 126.

50) **If a horse will probably win three races out of ten, what are the odds that he will win?**
(Skill 11.3)(Average Rigor)

A)   3:10

B)   7:10

C)   3:7

D)   7:3

Answer: C

The odds are that he will win 3 and lose 7.

51) **Given a drawer with 5 black socks, 3 blue socks, and 2 red socks, what is the probability that you will draw two black socks in two draws in a dark room?**
   (Skill 11.3)(Rigorous)

   A)   2/9

   B)   1/4

   C)   17/18

   D)   1/18

   Answer: A

   In this example of conditional probability, the probability of drawing a black sock on the first draw is 5/10. It is implied in the problem that there is no replacement, therefore the probability of obtaining a black sock in the second draw is 4/9. Multiply the two probabilities and reduce to lowest terms.

52) **Choose the least appropriate set of manipulatives for a sixth grade class.**
   (Skill 12.1) (Easy)

   A)   graphing calculators, compasses, rulers, conic section models

   B)   two color counters, origami paper, markers, yarn, balance, meter stick, colored pencils, beads

   C)   balance, meter stick, colored pencils, beads

   D)   paper cups, beans, tangrams, geoboards

   Answer: A

   The manipulatives in answer A include tools that are most appropriate for students studying more advanced topics in algebra, such as functions and conic sections, as well as more advanced topics in geometry. As a result, these manipulatives may not be appropriate for sixth grade material. The other answers include manipulatives that may be more appropriate for a sixth grade class.

53) **A teacher plans an activity that involves students calculating how many chair legs are in the classroom, given that there are 30 chairs and each chair has 4 legs. This activity is introducing the ideas of: (Skill 12.1)(Average rigor)**

A) Probability

B) Statistics

C) Geometry

D) Algebra

Answer: D

This activity involves recognizing patterns. It could also involve problem-solving by developing an expression that represents the problem. Activities such as this do not introduce the terms of algebra, but they introduce some of the ideas of algebra.

54) If 80 students are in sports and 100 students are in band with 20 students in both band and sports, identify that the appropriate pictorial representation.
(Skill 12.2) (Average Rigor)

A)

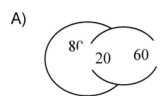

B)

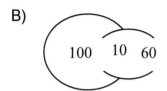

C)

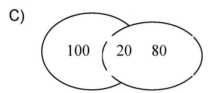

D)

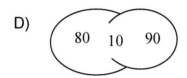

Answer: A

If 20 students are in both band and sports, you need to subtract twenty from those in just band and those in just sports, leaving the answer as A.

55) Tom's age is one less than half the sum of his brother's and sister's ages. Tom's brother is twice as old as he is and his sister is 2 years younger than Tom. This information can be expressed symbolically in the following way:
(Skill 12.3) (Rigorous)

A) $n = \frac{1}{2}(2n + n - 2) - 1$

B) $n = \frac{1}{2}(2n + n - 2) + 1$

C) $n = \frac{1}{2}(2n + n - 2 - 1)$

D) $n = \frac{1}{2}(2n + n - 2 + 1)$

Answer: A

56) What is the ratio of the frequencies of the third and sixth notes in a musical octave given that the frequency of a note = frequency of starting note $\times\ 2^{N/12}$ where *N* is the number of notes from the starting point.
(Skill 12.5) (Rigorous)

A) 1.19

B) 1.41

C) 0.71

D) 0.84

Answer: D

The frequency of the third note is $f_1 2^{\frac{3}{12}} = f_1 2^{\frac{1}{4}}$. The frequency of the sixth note is $f_1 2^{\frac{6}{12}} = f_1 2^{\frac{1}{2}}$. The ratio of the two frequencies is $2^{\frac{1}{4} - \frac{1}{2}} = 2^{-0.25} = 0.84$.

57)   **When you begin by assuming the conclusion of a theorem is false, then show that through a sequence of logically correct steps you contradict an accepted fact, this is known as (Easy)(Skill 13.1)**

    A)   inductive reasoning

    B)   direct proof

    C)   indirect proof

    D)   exhaustive proof

    Answer: C

    By definition this describes the procedure of an indirect proof.

58)   **Which of the following statements is logically equivalent to "It is not true that Gina owns no books" (Average Rigor)(Skill 13.3)**

    A) It is true that Gina owns a lot of books

    B) It is likely that Gina owns some books

    C) It is true that Gina owns books

    D) None of the above

    Answer: C

59) **Which of the following arguments is valid?**
(Rigorous)(Skill 13.3)

A) People who support tax reduction will vote for Mr.Allen. Mr.Brown voted for Mr.Allen. Conclusion: Mr. Brown supports tax reduction.

B) If Mr.Allen is elected, he will reduce taxes. Mr. Allen was elected. Conclusion: Taxes were reduced.

C) If Mr.Allen is elected, he will reduce taxes. Taxes were reduced. Conclusion: Mr. Allen was elected.

D) None of the above

Answer: B

A is not valid since people other than those who support tax reduction may vote for Mr. Allen. C is not valid since there may be another politician who will also reduce taxes.

60) **Which is the least appropriate strategy to emphasize when teaching problem solving? (Easy)(Skill 13.4)**

A)   guess and check

B)   look for key words to indicate operations such as all together—add, more than, subtract, times, multiply

C)   make a diagram

D)   solve a simpler version of the problem

Answer: B

Answers A, C and D are all legitimate methods for approaching a problem. Answer B, on the other hand, is much less general and not as broadly applicable as the other answers, and therefore constitutes the least appropriate strategy.

Sample Test II

1) How many real numbers lie between -1 and +1? (Skill 1.1) (Easy Rigor)

   A.   0

   B)   1

   C)   17

   D)   an infinite number

2) Which of the following is always composite if x is odd, y is even, and both x and y are greater than or equal to 2? (Skill 1.2)(Average Rigor)

   A) $x + y$

   B) $3x + 2y$

   C) $5xy$

   D) $5x + 3y$

3) Given that n is a positive even integer, $5n + 4$ will always be divisible by: (Skill 1.2)(Average Rigor)

   A)   4

   B)   5

   C) $5n$

   D)   2

4) What is the Greatest Common Factor of 25 and 40? (Skill 1.4) (Average Rigor)

   A)   10

   B)   5

   C)   8

   D)   1

5) Given even numbers x and y, which is the LCM of x and y? (Skill 1.4) (Average Rigor)

   A) $\dfrac{xy}{2}$

   B) $2xy$

   C) $4xy$

   D) xy

6) Choose the set in which the members are <u>not</u> equivalent. (Skill 2.2)(Easy Rigor)

   A)   1/2, 0.5, 50%

   B)   10/5, 2.0, 200%

   C)   3/8, 0.385, 38.5%

   D)   7/10, 0.7, 70%

7) Change $.\overline{63}$ into a fraction in simplest form. (Skill 2.2)(Easy Rigor)

A) $63/100$

B) $7/11$

C) $6\ 3/10$

D) $2/3$

8) Write the number 81 in exponent form. (Skill 2.2)(Average Rigor)

A) 3^4

B) 2^5

C) 3^3

D) 3^5

9) Which is not true? (Skill 2.3)(Easy Rigor)

A) All irrational numbers are real numbers.

B) All integers are rational.

C) Zero is a natural number.

D) All whole numbers are integers.

10) Mr. Brown feeds his cat premium cat food which costs $40 per month. Approximately how much will it cost to feed the cat for one year? (Skill 2.4)(Average Rigor)

A) $500

B) $400

C) $80

D) $4,800

11) What would be the total cost of a suit for $295.99 and a pair of shoes for $69.95 including 6.5% sales tax? (Skill 2.5) (Average Rigor)

A) $389.73

B) $398.37

C) $237.86

D) $315.23

12) Which statement is an example of the Identity Property of Addition? (Skill 3.1)(Easy Rigor)

A) $3 + -3 = 0$

B) $3x = 3x + 0$

C) $3 \times \dfrac{1}{3} = 1$

D) $3 + 2x = 2x + 3$

13) Simplify. $(3.8 \times 10^{17}) \times (.5 \times 10^{-12})$
(Skill 3.3) (Rigorous)

A) 19×10^5

B) 1.9×10^5

C) 1.9×10^6

D) 1.9×10^7

14) If cleaning costs are $32 for 4 hours, how much is it for 10.5 hours of cleaning?
(Skill 3.4)(Average Rigor)

A) $112.50

B) $87

C) $84

D) $76.50

15) 3 km is equivalent to
(Skill 4.1)(Easy Rigor)

A)   300 cm

B)   300 m

C)   3,000 cm

D)   3,000 m

16)

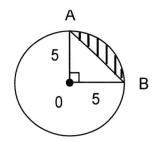

Compute the area of the shaded region of he circle, given a radius of 5 meters. Point *0* is the center.
(Skill 4.4)(Rigorous)

A)   7.13 cm²

B)   7.13 m²

C)   78.5 m²

D)   19.63 m²

17) Find the area of the figure pictured below.
(Skill 4.4)(Rigorous)

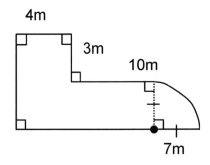

A)   136.47 m²

B)   148.48 m²

C)   293.86 m²

D)   178.47 m²

18) A 30 meter x 60 meter rectangular garden contains a circular fountain with a 5 meter radius. Calculate the area of the portion of the garden not occupied by the fountain.
(Skill 4.4)(Rigorous)

A) 1721 m²

B) 1879 m²

C) 2585 m²

D) 1015 m²

19) If the area of the base of a cone is tripled, the volume will be
(Skill 4.6)(Average Rigor)

A) the same as the original

B) 9 times the original

C) 3 times the original

D) 3π times the original

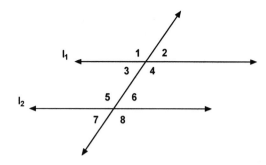

20) Given $l_1 \parallel l_2$ (parallel lines 1 & 2) which of the following is true?
(Skill 5.1)(Average Rigor)

A) $\angle 1$ and $\angle 8$ are congruent and alternate interior angles

B) $\angle 2$ and $\angle 3$ are congruent and corresponding angles

C) $\angle 3$ and $\angle 4$ are adjacent and supplementary angles

D) $\angle 3$ and $\angle 5$ are adjacent and supplementary angles

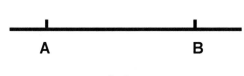

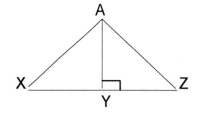

23) Given $XY \cong YZ$ and $\angle AYX \cong \angle AYZ$. Prove $\triangle AYZ \cong \triangle AYX$.

21) The above construction can be completed to make (Skill 5.2)(Average Rigor)

A)   an angle bisector

B)   parallel lines

C)   a perpendicular bisector

D)   skew lines

1) $XY \cong YZ$

2) $\angle AYX \cong \angle AYZ$

3) $AY \cong AY$

4) $\triangle AYZ \cong \triangle AYX$

Which property justifies step 3? (Skill 5.3)(Rigorous)

A)   reflexive

B)   symmetric

C)   transitive

D)   identity

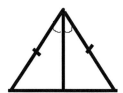

22) What method could be used to prove the above triangles congruent? (Skill 5.3)(Average Rigor)

A)   SSS

B)   SAS

C)   AAS

D)   SSA

24) What is the degree measure of an interior angle of a regular 10-sided polygon? (Skill 5.4)(Average Rigor)

A)   18°

B)   36°

C)   144°

D)   54°

25) A car is driven north at 74 miles per hour from point A. Another car is driven due east at 65 miles per hour starting from the same point at the same time. How far are the cars away from each other after 2 hours?
(Skill 5.5) (Rigorous)

A) 175.87 miles

B) 232.66 miles

C) 196.99 miles

D) 202.43 miles

26) Find the midpoint of a segment with endpoints at (2, 5) and (7, -4).
(Skill 6.2)(Rigorous)

A) (9,-1)

B) (5,9)

C) (9/2 , -1/2)

D) (9/2, 1/2)

27) Find the slope of the line that crosses points (25, 11) and (4, 4).
(Skill 6.2)(Rigorous)

A) 1/3

B) 4

C) 4.5

D) 1.7

28) Given similar polygons with corresponding sides measuring 6 and 8, what is the area of the smaller polygon if the area of the larger is 64?
(Skill 6.4)(Average Rigor)

A) 48

B) 36

C) 144

D) 78

29) {6, 11, 16, 21, . .}
Find the sum of the first 20 terms in the sequence.
(Skill 7.1)(Rigorous)

A) 1,070

B) 1,176

C) 969

D) 1,069

30) **Give the domain for the function over the set of real numbers:**

$$\frac{3x+2}{2x^2-3}$$

(Skill 7.3)(Rigorous)

A)   all real numbers

B)   all real numbers, $x \neq 0$

C)   all real numbers, $x \neq -2$ or 3

D)   all real numbers, $x \neq \dfrac{\pm\sqrt{6}}{2}$

31) If *y* varies directly with *x*, and *x* = 2 when *y* = 6, what is *x* when *y* = 18?
(Skill 7.4)(Rigorous)

A)   3

B)   6

C)   26

D)   36

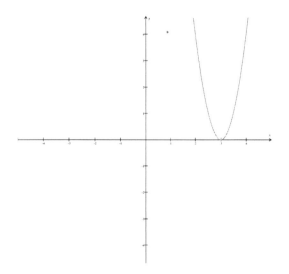

32) Which equation is graphed above?
(Skill 7.5)(Rigorous)

A) $y = 4(x + 3)^2$

B) $y = 4(x - 3)^2$

C) $y = 3(x - 4)^2$

D) $y = 3(x + 4)^2$

33) Which graph represents the equation of
$y = x^2 + 3x$?
(Skill 7.5)(Rigorous)

A)   B)

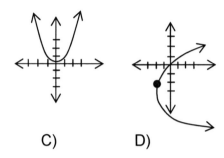

C)   D)

34) Simplify $\dfrac{\frac{3}{4}x^2y^{-3}}{\frac{2}{3}xy}$
(Skill 8.1)(Average Rigor)

A) $\frac{1}{2}xy^{-4}$

B) $\frac{1}{2}x^{-1}y^{-4}$

C) $\frac{9}{8}xy^{-4}$

D) $\frac{9}{8}xy^{-2}$

35) Factor completely.
$8(x - y) + a(y - x)$
(Skill 8.1)(Rigorous)

   A) $(8 + a)(y - x)$

   B) $(8 - a)(y - x)$

   C) $(a - 8)(y - x)$

   D) $(a - 8)(y + x)$

36) Which of the following is a factor of $6 + 48m^3$
(Skill 8.1)(Rigorous)

   A) $(1 + 2m)$

   B) $(1 - 8m)$

   C) $(1 + m - 2m)$

   D) $(1 - m + 2m)$

37) Solve for x:
$|2x + 3| > 4$
(Skill 8.3)(Rigorous)

   A) $-\frac{7}{2} > x > \frac{1}{2}$

   B) $-\frac{1}{2} > x > \frac{7}{2}$

   C) $x < \frac{7}{2}$ or $x < -\frac{1}{2}$

   D) $x < -\frac{7}{2}$ or $x > \frac{1}{2}$

38) Solve for x: $\dfrac{4}{x} = \dfrac{8}{3}$
(Skill 8.5)(Easy Rigor)

   A) $x = 0.66666...$

   B) $x = 0.6$

   C) $x = 15$

   D) $x = 1.5$

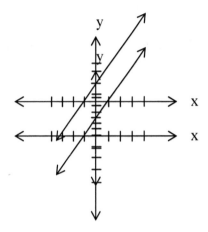

39) What is the equation of the above graph?
(Skill 9.2)(Rigorous)

   A) $2x + y = 2$

   B) $2x - y = -2$

   C) $2x - y = 2$

   D) $2x + y = -2$

40) Solve the system of equations.

$x = 3y + 7$
$7x + 5y = 23$
(Skill 9.5)(Average Rigor)

A) (-1, 4)

B) (4, -1)

C) $(\frac{-29}{7}, \frac{-26}{7})$

D) (10, 1)

41) Solve the following system of equations: $4x + 3y = 24$ and $3x + 2y = 20$.
(Skill 9.5)(Rigorous)

A) (4/3, 3/2)

B) (4, 5)

C) (5, 5)

D) (12, - 8)

42) Which graph represents the solution set for
$x^2 - 5x > -6$?
(Skill 9.7)(Rigorous)

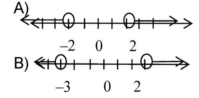

A)
$-2 \quad 0 \quad 2$

B)
$-3 \quad 0 \quad 2$

C)
$-2 \quad 0 \quad 2$

D)
$-3 \quad 0 \quad 2 \quad 3$

43) Which of the following is not a valid method of collecting statistical data?
(Average Rigor)(Skill10.1)

A) Random sampling

B) Systematic sampling

C) Cluster sampling

D) Cylindrical sampling

44) Which of the following types of graphs would be best to display the eye color of the students in a class?
(Average rigor) (Skill 10.2)

A) Bar graph or circle graph

B) Pictograph or bar graph

C) Line graph or pictograph

D) Line graph or bar graph

45) What conclusion can be drawn from the graph below? (Skill 10.3)(Easy Rigor)

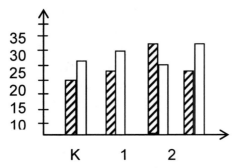

MLK Elementary Student Enrollment   ▨ Girls   ☐ Boys

A)   The number of students in first grade exceeds the number in second grade.

B)   There are more boys than girls in the entire school.

C)   There are more girls than boys in the first grade.

D)   Third grade has the greatest number of students.

46) What is the first, second and third quartile for the following? (Skill 10.4)(Average Rigor)

5, 5, 5, 6, 7, 9, 9, 10, 11, 12, 13, 13, 14, 15, 16, 17, 17

A)   5, 10, 15

B)   6, 11, 16

C)   6, 11, 14

D)   7, 11, 14

47) A student scored in the 87th percentile on a standardized test. Which would be the best interpretation of his score? (Skill 10.4)(Average Rigor)

A)   Only 13% of the students who took the test scored higher.

B)   This student should be getting mostly B's on his report card.

C)   This student performed below average on the test.

D)   This is the equivalent of missing 13 questions on a 100-question exam.

48) A measure of association between two variables is called:
(Skill 10.6) (Easy Rigor)

A) Associate

B) Correlation

C) Confidence interval

D) Variation

49) Determine the number of subsets of set K.
$K = \{4, 5, 6, 7\}$
(Skill 11.1)(Average Rigor)

A) 15

B) 16

C) 17

D) 18

50) Given a spinner with the numbers one through eight, what is the probability that you will spin an even number or a number greater than four?
(Skill 11.3)(Average Rigor)

A) ¼

B) ½

C) ¾

D) 1

51) A sack of candy has 3 peppermints, 2 butterscotch drops and 3 cinnamon drops. One candy is drawn and replaced, then another candy is drawn; what is the probability that both will be butterscotch?
(Skill 11.3)(Rigorous)

A) 1/2

B) 1/28

C) 1/4

D) 1/16

52) A teacher is introducing the concept of multiplication to her third grade students. What is another way she might write 4 x 5?
(Skill 12.1)(Average rigor)

A) 4 + 5

B) 5 + 4

C) 4 + 4 + 4 + 4 + 4

D) 5 + 5 + 5 + 5 + 5

53) What math principle is reinforced by matching numerals with number words?
(Skill 12.1)(Rigorous)

    A) Sequencing

    B) Greater than and less than

    C) Number representations

    D) Rote counting

54) Line p has a negative slope and passes through the point $(0, 0)$. If line q is perpendicular to line p, which of the following must be true?
(Skill 12.2) (Rigorous)

    A) Line q has a negative y-intercept.

    B) Line q passes through the point $(0, 0)$

    C) Line q has a positive slope.

    D) Line q has a positive y-intercept.

55) Cindy bought a package of cookies. She ate half of them and gave one-third of the remainder to a friend. She then had 20 fewer cookies than she did in the beginning. This information can be expressed symbolically in the following way:
(Skill 12.3) (Rigorous)

    A) $\dfrac{n}{2} - \dfrac{n}{3} = n - 20$

    B) $\dfrac{n}{2} - \dfrac{n}{6} = n - 20$

    C) $\dfrac{n}{2}(1 - \dfrac{1}{3}) = n - 20$

    D) $\dfrac{n}{2} - \dfrac{n}{3} = n + 20$

56) If n is a multiple of 3, then $n + 5$ is divisible by 4.

$n = 6$ is a multiple of 3.
$n + 5 = 6 + 5 = 11$.
11 is not divisible by 4.

The above is an example of:
(Skill 13.1) (Average Rigor)

    A) Direct proof

    B) Indirect proof

    C) Counter example

    D) None of the above

57) Which of the following is an example of syllogism? (Skill 13.3) (Rigorous)

A) When it rains he carries an umbrella. It did not rain and so he did not carry an umbrella.

B) Parrots are birds. Eagles are birds. Therefore parrots are eagles.

C) All vegetables are nutritious. Yams are vegetables. Therefore yams are nutritious.

D) None of the above

58) A student turns in a paper with this type of error:
$7 + 16 \div 8 \times 2 = 8$
$8 - 3 \times 3 + 4 = -5$
(Easy)(Skill 13.4)

In order to remediate this error, a teacher should:

A) review and drill basic number facts

B) emphasize the importance of using parentheses in simplifying expressions

C) emphasize the importance of working from left to right when applying the order of operations

D) do nothing; these answers are correct

59) **Identify the proper sequencing of subskills when teaching graphing inequalities in two dimensions (Easy)(Skill 13.4)**

A) shading regions, graphing lines, graphing points, determining whether a line is solid or broken

B) graphing points, graphing lines, determining whether a line is solid or broken, shading regions

C) graphing points, shading regions, determining whether a line is solid or broken, graphing lines

D) graphing lines, determining whether a line is solid or broken, graphing points, shading regions

60) **Which statement is incorrect? (Easy)(Skill 13.5)**

A) Drill and practice is one good use for classroom computers.

B) Computer programs can help to teach problem solving in the classroom.

C) Computers are not effective unless each child in the class has his or her own workstation.

D) Analyzing science project data on a computer during math class is an excellent use of class time.

Sample Test II - Answer Key

1. D		44. B
2. C		45. B
3. D		46. D
4. B		47. A
5. A		48. B
6. C		49. B
7. B		50. C
8. A		51. D
9. C		52. C
10. A		53. C
11. A		54. C
12. B		55. B
13. B		56. C
14. C		57. C
15. D		58. C
16. B		59. B
17. B		60. C
18. A		
19. C		
20. C		
21. C		
22. B		
23. A		
24. C		
25. C		
26. D		
27. A		
28. B		
29. A		
30. D		
31. B		
32. B		
33. C		
34. C		
35. C		
36. A		
37. D		
38. D		
39. B		
40. B		
41. D		
42. D		
43. D		

Sample Test II - Rigor Table

Easy Rigor 20%	Average Rigor 40%	Rigorous 40%
1, 6, 7, 9, 12, 15, 38, 45, 48, 58, 59, 60	2, 3, 4, 5, 8, 10, 11, 14, 19, 20, 21, 22, 24, 28, 34, 40, 43, 44, 46, 47, 49, 50, 52, 56	13, 16, 17, 18, 23, 25, 26, 27, 29, 30, 31, 32, 33, 35, 36, 37, 39, 41, 42, 51, 53, 54, 55, 57

Rationales for Sample Test II

1) **How many real numbers lie between -1 and +1? (Skill 1.1) (Easy Rigor)**

    A)   0

    B)   1

    C)   17

    D)   an infinite number

    Answer: D

    There are an infinite number of real numbers between any two real numbers.

2) **Which of the following is always composite if *x* is odd, *y* is even, and both *x* and *y* are greater than or equal to 2? (Skill 1.2)(Average Rigor)**

    A) $x + y$

    B) $3x + 2y$

    C) $5xy$

    D) $5x + 3y$

    Answer: C

    A composite number is a number that is not prime. The prime number sequence begins 2, 3, 5, 7, 11, 13, 17,....
    To determine which of the expressions is <u>always</u> composite, experiment with different values of *x* and *y*, such as *x* = 3 and *y* = 2, or *x* = 5 and *y* = 2. It turns out that 5*xy* will always be an even number and therefore composite since *y* is an even number.

3)   Given that *n* is a positive even integer, *5n* + 4 will always be divisible by:
(Skill 1.2)(Average Rigor)

A)   4

B)   5

C)   *5n*

D)   2

Answer: D

$5n$ is always even and an even number added to an even number is always an even number, thus divisible by 2.

4)   What is the Greatest Common Factor of 25 and 40?
(Skill 1.4) (Average Rigor)

A)   10

B)   5

C)   8

D)   1

Answer: B

In terms of prime factors, $25 = 5^2$ and $40 = 2^3 \times 5$. The greatest common factor is 5.

5) **Given even numbers x and y, which is the LCM of *x* and *y*?**
   (Skill 1.4) (Average Rigor)

   A) $\dfrac{xy}{2}$

   B) $2xy$

   C) $4xy$

   D) xy

   Answer: A

   Although choices B, C and D are common multiples, when both numbers are even, the product can be divided by two to obtain the least common multiple.

6) **Choose the set in which the members are <u>not</u> equivalent.**
   (Skill 2.2)(Easy Rigor)

   A) 1/2, 0.5, 50%

   B) 10/5, 2.0, 200%

   C) 3/8, 0.385, 38.5%

   D) 7/10, 0.7, 70%

   Answer: C

   3/8 is equivalent to .375 and 37.5%

7)   Change $.\overline{63}$ into a fraction in simplest form. (Skill 2.2)(Easy Rigor)

     A) $63/100$

     B) $7/11$

     C) $6\ 3/10$

     D) $2/3$

Answer: B

Let N = .636363… Then multiplying both sides of the equation by 100 or 10^2 (because there are 2 repeated numbers), we get 100N = 63.636363… Then subtracting the two equations (N = .636363… and 100N = 63.636363…), gives 99N = 63 or N = $\dfrac{63}{99}=\dfrac{7}{11}$.

8)   Write the number 81 in exponent form.
     (Skill 2.2)(Average Rigor)

     A) 3^4

     B) 2^5

     C) 3^3

     D) 3^5

Answer: A

81 = 3 x 3 x 3 x 3 or 3^4.

9) **Which is not true?**
   (Skill 2.3)(Easy Rigor)

   A)   All irrational numbers are real numbers.

   B)   All integers are rational.

   C)   Zero is a natural number.

   D)   All whole numbers are integers.

   Answer: C

   Zero is not a natural number or "counting number"

10) **Mr. Brown feeds his cat premium cat food which costs $40 per month. Approximately how much will it cost to feed the cat for one year?**
    (Skill 2.4)(Average Rigor)

    A) $500

    B) $400

    C) $80

    D) $4,800

    Answer: A

    12(40) = 480 which is closest to $500

11) **What would be the total cost of a suit for $295.99 and a pair of shoes for $69.95 including 6.5% sales tax?**
(Skill 2.5) (Average Rigor)

A) $389.73

B) $398.37

C) $237.86

D) $315.23

Answer: A

Before the tax, the total comes to $365.94. Then .065(365.94) = 23.79. With the tax added on, the total bill is 365.94 + 23.79 = $389.73. (Quicker way: 1.065(365.94) = 389.73.)

12) **Which statement is an example of the identity Property of Addition?**
(Skill 3.1)(Easy Rigor)

A) $3 + -3 = 0$

B) $3x = 3x + 0$

C) $3 \times \dfrac{1}{3} = 1$

D) $3 + 2x = 2x + 3$

Answer: B

A illustrates additive inverse, C illustrates the multiplicative inverse, and D illustrates the Commutative Property of Addition.

13) **Simplify. $(3.8 \times 10^{17}) \times (.5 \times 10^{-12})$**
(Skill 3.3) (Rigorous)

A) 19×10^5

B) 1.9×10^5

C) 1.9×10^6

D) 1.9×10^7

Answer: B

Multiply the decimals and add the exponents.

14) **If cleaning costs are $32 for 4 hours, how much is it for 10.5 hours of cleaning?**
(Skill 3.4)(Average Rigor)

A) $112.50

B) $87

C) $84

D) $76.50

Answer: C

The hourly rate is $8 per hour so 8 x 10.5 = $84.

15) **3 km is equivalent to**
(Skill 4.1)(Easy Rigor)

A) 300 cm

B) 300 m

C) 3,000 cm

D) 3,000 m

Answer: D

1 kilometer = 1,000 meters.

16)

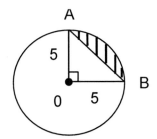

Compute the area of the shaded region of the circle, given a radius of 5 meters. Point *0* is the center.
(Skill 4.4)(Rigorous)

A)   7.13 cm²

B)   7.13 m²

C)   78.5 m²

D)   19.63 m²

Answer: B

Area of triangle *AOB* is .5(5)(5) = 12.5 square meters. Since $\frac{90}{360} = .25$, the area of sector *AOB* (pie-shaped piece) is approximately .25(π)5² = 19.63. Subtracting the triangle area from the sector area to get the area of the shaded region, we get approximately 19.63 – 12.5 = 7.13 square meters.

17) **Find the area of the figure pictured below.**
(Skill 4.4)(Rigorous)

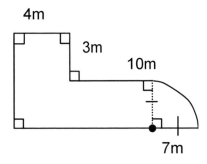

A) 136.47 m²

B) 148.48 m²

C) 293.86 m²

D) 178.47 m²

Answer: B

Divide the figure into 2 rectangles and one quarter circle. The tall rectangle on the left will have dimensions 10 by 4 and area 40. The rectangle in the center will have dimensions 7 by 10 and area 70. The quarter circle will have area $.25(\pi)7^2 = 38.48$. The total area is therefore approximately 148.48.

18) **A 30 meter x 60 meter rectangular garden contains a circular fountain with a 5 meter radius. Calculate the area of the portion of the garden not occupied by the fountain.**
(Skill 4.4)(Rigorous)

A) 1721 m²

B) 1879 m²

C) 2585 m²

D) 1015 m²

Answer: A

Find the area of the garden and then subtract the area of the fountain: $30(60) - \pi(5)^2$ or approximately 1721 square meters.

19) **If the area of the base of a cone is tripled, the volume will be (Skill 4.6)(Average Rigor)**

    A)   the same as the original

    B)   9 times the original

    C)   3 times the original

    D) 3π times the original

Answer: C

The formula for the volume of a cone is $V = \frac{1}{3}Bh$, where B is the area of the circular base and h is the height. If the area of the base is tripled, the volume becomes $V = \frac{1}{3}(3B)h = Bh$, or three times the original area.

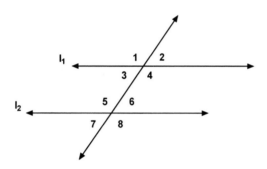

20) **Given $l_1 \parallel l_2$ (parallel lines 1 & 2) which of the following is true? (Skill 5.1)(Average Rigor)**

    A) $\angle 1$ and $\angle 8$ are congruent and alternate interior angles

    B) $\angle 2$ and $\angle 3$ are congruent and corresponding angles

    C) $\angle 3$ and $\angle 4$ are adjacent and supplementary angles

    D) $\angle 3$ and $\angle 5$ are adjacent and supplementary angles

Answer: C

The angles in choice A are alternate exterior angles. In choice B, the angles are vertical. The angles in choice D are consecutive, not adjacent.

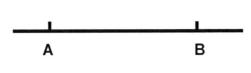

21) **The above construction can be completed to make (Skill 5.2)(Average Rigor)**

A)   an angle bisector

B)   parallel lines

C)   a perpendicular bisector

D)   skew lines

Answer: C

The points marked C and D are the intersection of the circles with centers A and B.

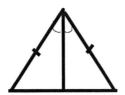

22) **What method could be used to prove the above triangles congruent? (Skill 5.3)(Average Rigor)**

A) SSS

B) SAS

C) AAS

D) SSA

Answer: B

Use SAS with the last side being the vertical line common to both triangles.

23) Given $XY \cong YZ$ and $\angle AYX \cong \angle AYZ$. Prove $\triangle AYZ \cong \triangle AYX$.

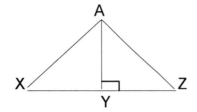

1) $XY \cong YZ$

2) $\angle AYX \cong \angle AYZ$

3) $AY \cong AY$

4) $\triangle AYZ \cong \triangle AYX$

Which property justifies step 3?
(Skill 5.3)(Rigorous)

A) reflexive

B) symmetric

C) transitive

D) identity

Answer: A

The reflexive property states that every number or variable is equal to itself and every segment is congruent to itself.

24) **What is the degree measure of an interior angle of a regular 10-sided polygon?**
(Skill 5.4)(Average Rigor)

A)   18°

B)   36°

C)   144°

D)   54°

Answer: C

Formula for finding the measure of each interior angle of a regular polygon with n sides is $\frac{(n-2)180}{n}$. For $n = 10$, we get $\frac{8(180)}{10} = 144$.

25) **A car is driven north at 74 miles per hour from point A. Another car is driven due east at 65 miles per hour starting from the same point at the same time. How far are the cars away from each other after 2 hours?**
(Skill 5.5) (Rigorous)

A)   175.87 miles

B)   232.66 miles

C)   196.99 miles

D)   202.43 miles

Answer: C

The route the cars take form a right triangle with edges 74 x 2 and 65 x 2. This gives two sides of a right triangle of 148 and 130. Using the Pythagorean Theorem, we get $148^2 + 130^2 = \text{distance}^2$. Therefore, distance between the cars = 196.99.

26) **Find the midpoint of a segment with endpoints at (2, 5) and (7, -4). (Skill 6.2)(Rigorous)**

    A)   (9,-1)

    B)   (5,9)

    C)   (9/2 , -1/2)

    D)   (9/2, 1/2)

Answer: D

Use the midpoint formula
$x = (2 + 7)/2 \quad y = (5 + -4)/2$

27) **Find the slope of the line that crosses points (25, 11) and (4, 4). (Skill 6.2)(Rigorous)**

    A)   1/3

    B)   4

    C)   4.5

    D)   1.7

Answer: A

Apply the formula $\dfrac{y_2 - y_1}{x_2 - x_1} = \dfrac{4 - 11}{4 - 25} = \dfrac{-7}{-21} = \dfrac{1}{3}$

28) **Given similar polygons with corresponding sides measuring 6 and 8, what is the area of the smaller polygon if the area of the larger is 64? (Skill 6.4)(Average Rigor)**

   A)   48

   B)   36

   C)   144

   D)   78

   Answer: B

   In similar polygons, the areas are proportional to the squares of the sides.
   $36/64 = x/64$

29) **{6, 11, 16, 21, . .}**
   Find the sum of the first 20 terms in the sequence.
   (Skill 7.1)(Rigorous)

   A)   1,070

   B)   1,176

   C)   9,69

   D)   1,069

   Answer: A

   Apply the formula $\frac{n}{2}[2a_1 + (n-1)d]$ where $n = 20$, $a_1 = 6$ and $d = 5$.

30) **Give the domain for the function over the set of real numbers:**

$$\frac{3x + 2}{2x^2 - 3}$$

(Skill 7.3)(Rigorous)

A) all real numbers

B) all real numbers, $x \neq 0$

C) all real numbers, $x \neq -2$ or 3

D) all real numbers, $x \neq \dfrac{\pm\sqrt{6}}{2}$

Answer: D

Solve the denominator for 0. These values will be excluded from the domain.

$$2x^2 - 3 = 0$$

$$2x^2 = 3$$

$$x^2 = 3/2$$

$$x = \sqrt{\tfrac{3}{2}} = \sqrt{\tfrac{3}{2}} \bullet \sqrt{\tfrac{2}{2}} = \tfrac{\pm\sqrt{6}}{2}$$

31) **If y varies directly with x, and $x = 2$ when $y = 6$, what is x when $y = 18$?**
(Skill 7.4)(Rigorous)

A) 3

B) 6

C) 26

D) 36

Answer: B

The equation for direct variation is $y = kx$ or $k = \dfrac{y}{x} = \dfrac{6}{2} = 3$. Substitute 18 for y and 3 for k and solve: $18 = 3x$, $x = 6$.

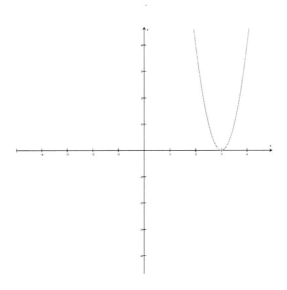

32) **Which equation is graphed above?**
(Skill 7.5)(Rigorous)

    A) $y = 4(x + 3)^2$

    B) $y = 4(x - 3)^2$

    C) $y = 3(x - 4)^2$

    D) $y = 3(x + 4)^2$

Answer: B

Since the vertex of the parabola is three units to the right of the origin, we choose the solution where 3 is subtracted from x, then the quantity is squared.

33) **Which graph represents the equation of** $y = x^2 + 3x$ **?**
(Skill 7.5)(Rigorous)

A) B)

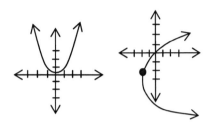

C) D)

Answer: C

Choice B is not the graph of a function. Choice D is the graph of a parabola where the coefficient of x^2 is negative. Choice A appears to be the graph of $y = x^2$. To find the x-intercepts of $y = x^2 + 3x$, set $y = 0$ and solve for x: $0 = x^2 + 3x = x(x + 3)$ to get $x = -3$. Therefore, the graph of the function intersects the x-axis at $x = -3$.

34) Simplify $\dfrac{\frac{3}{4}x^2y^3}{\frac{2}{3}xy}$

(Skill 8.1)(Average Rigor)

A) $\frac{1}{2}xy^{-4}$

B) $\frac{1}{2}x^{-1}y^{-4}$

C) $\frac{9}{8}xy^{-4}$

D) $\frac{9}{8}xy^{-2}$

Answer: C

Simplify the complex fraction by inverting the denominator and multiplying: 3/4(3/2) = 9/8, then subtract exponents to obtain the correct answer.

35) **Factor completely.**
$8(x - y) + a(y - x)$
(Skill 8.1)(Rigorous)

A) $(8 + a)(y - x)$

B) $(8 - a)(y - x)$

C) $(a - 8)(y - x)$

D) $(a - 8)(y + x)$

Answer: C

Glancing first at the solution choices, factor $(y - x)$ from each term. This leaves -8 from the first reran and a from the send term: $(a - 8)(y - x)$.

36) **Which of the following is a factor of** $6 + 48m^3$
   (Skill 8.1)(Rigorous)

   A) $(1 + 2m)$

   B) $(1 - 8m)$

   C) $(1 + m - 2m)$

   D) $(1 - m + 2m)$

   Answer: A

   Removing the common factor of 6 and then factoring the sum of two cubes gives $6 + 48m^3 = 6(1 + 8m^{3)} = 6(1 + 2m)(1^2 - 2m + (2m)^2)$.

37) **Solve for *x*:**
   $|2x + 3| > 4$
   (Skill 8.3)(Rigorous)

   A) $-\frac{7}{2} > x > \frac{1}{2}$

   B) $-\frac{1}{2} > x > \frac{7}{2}$

   C) $x < \frac{7}{2}$ or $x < -\frac{1}{2}$

   D) $x < -\frac{7}{2}$ or $x > \frac{1}{2}$

   Answer: D

   The quantity within the absolute value symbols must be either > 4 or < -4. Solve the two inequalities $2x + 3 > 4$ and $2x + 3 < -4$.

38) **Solve for x:** $\dfrac{4}{x} = \dfrac{8}{3}$

(Skill 8.5)(Easy Rigor)

A) $x = 0.66666...$

B) $x = 0.6$

C) $x = 15$

D) $x = 1.5$

Answer: D

Cross multiply to obtain $12 = 8x$, then divide both sides by 8.

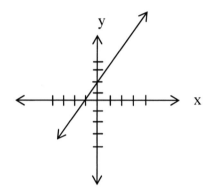

39) **What is the equation of the graph below?**
(Skill 9.2)(Rigorous)

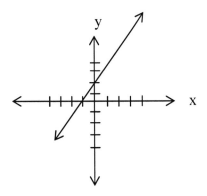

A) $2x + y = 2$

B) $2x - y = -2$

C) $2x - y = 2$

D) $2x + y = -2$

Answer: B

By observation, we see that the graph has a y-intercept of 2 and a slope of 2. Therefore its equation is $y = mx + b = 2x + 2$. Rearranging the terms gives $2x - y = -2$.

40) **Solve the system of equations.**

$x = 3y + 7$
$7x + 5y = 23$

(Skill 9.5)(Average Rigor)

A) $(-1, 4)$

B) $(4, -1)$

C) $(\frac{-29}{7}, \frac{-26}{7})$

D) $(10, 1)$

Answer: B

Substituting x in the second equation results in $7(3y + 7)+ 5y = 23$.
Solve by distributing and grouping like terms:
$26y+49 = 23$, $26y = -26$, $y = -1$
Substitute y into the first equation to obtain x.

41) **Solve the following system of equations: $4x + 3y = 24$ and $3x + 2y = 20$.**
(Skill 9.5)(Rigorous)

A) $(4/3, 3/2)$

B) $(4, 5)$

C) $(5, 5)$

D) $(12, -8)$

Answer: D

Solving the first equation for x we get $x = \dfrac{24 - 3y}{4}$. Placing that value in

the second equation we have:

$$\frac{3(24 - 3y)}{4} + 2y = 20$$
$$3(24 - 2y) + 8y = 80$$
$$-y = 8$$
$$y = -8$$

Substituting the value of y in $x = \dfrac{24 - 3y}{4}$, we get $x = 12$.

42) **Which graph represents the solution set for**
$x^2 - 5x > -6$ **?**
(Skill 9.7)(Rigorous)

A)
$-2 \quad 0 \quad 2$

B)
$-3 \quad 0 \quad 2$

C)
$-2 \quad 0 \quad 2$

D)
$-3 \quad 0 \quad 2 \; 3$

Answer: D

Rewriting the inequality gives $x^2 - 5x + 6 > 0$. Factoring gives $(x - 2)(x - 3) > 0$. The two cut-off points on the number line are now at $x = 2$ and $x = 3$. Choosing a random number in each of the three parts of the number line, we test them to see if they produce a true statement. If $x = 0$ or $x = 4$, $(x - 2)(x - 3) > 0$ is true. If $x = 2.5$, $(x - 2)(x - 3) > 0$ is false. Therefore the solution set is all numbers less than 2 and greater than 3.

43) **Which of the following is not a valid method of collecting statistical data?**
(Average Rigor)(Skill10.1)

   A)   Random sampling

   B)   Systematic sampling

   C)   Cluster sampling

   D)   Cylindrical sampling

Answer: D

There is no such method as cylindrical sampling.

44) **Which of the following types of graphs would be best to display the eye color of the students in a class?**
(Average rigor) (Skill 10.2)

A) Bar graph or circle graph

B) Pictograph or bar graph

C) Line graph or pictograph

D) Line graph or bar graph

Answer: B

A pictograph or a line graph could be used. In this activity, a line graph would not be used because it shows change over time. Although a circle graph could be used to show a percentage of students with brown eyes, blue eyes, etc. that representation would be too advanced for early childhood students.

45) **What conclusion can be drawn from the graph below?**
(Skill 10.3)(Easy Rigor)

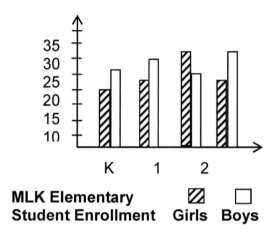

MLK Elementary
Student Enrollment   Girls Boys

A)   The number of students in first grade exceeds the number in second grade.

B)   There are more boys than girls in the entire school.

C)   There are more girls than boys in the first grade.

D)   Third grade has the greatest number of students.

Answer: B

In Kindergarten, first grade, and third grade, there are more boys than girls. The number of extra girls in grade two is more than made up for by the extra boys in all the other grades put together.

46) **What is the first, second and third quartile for the following?**
(Skill 10.4)(Average Rigor)

5, 5, 5, 6, 7, 9, 9, 10, 11, 12, 13, 13, 14, 15, 16, 17, 17

A)   5, 10, 15

B)   6, 11, 16

C)   6, 11, 14

D)   7, 11, 14

Answer: D

47) **A student scored in the 87th percentile on a standardized test.**
Which would be the best interpretation of his score?
(Skill 10.4)(Average Rigor)

A)   Only 13% of the students who took the test scored higher.

B)   This student should be getting mostly B's on his report card.

C)   This student performed below average on the test.

D)   This is the equivalent of missing 13 questions on a 100-question exam.

Answer: A

Percentile ranking tells how the student compared to the norm or the other students taking the test. It does not correspond to the percentage answered correctly, but can indicate how the student compared to the average student tested.

48) **A measure of association between two variables is called:**
(Skill 10.6) (Easy Rigor)

A)   Associate

B)   Correlation

C)   Confidence interval

D)   Variation

Answer: B

49) **Determine the number of subsets of set K.**
K = {4, 5, 6, 7}
(Skill 11.1)(Average Rigor)

A)   15

B)   16

C)   17

D)   18

Answer: B

A set of n objects has 2^n subsets. Therefore, here we have $2^4 = 16$ subsets. These subsets include four which each have 1 element only, six which each have 2 elements, four which each have 3 elements, plus the original set, and the empty set.

50) **Given a spinner with the numbers one through eight, what is the probability that you will spin an even number or a number greater than four?**
(Skill 11.3)(Average Rigor)

A)   ¼

B)   ½

C)   ¾

D)   1

Answer: C

There are 6 favorable outcomes: 2, 4, 5, 6, 7, 8 and 8 possibilities. Reduce 6/8 to 3/4.

51) **A sack of candy has 3 peppermints, 2 butterscotch drops and 3 cinnamon drops. One candy is drawn and replaced, then another candy is drawn; what is the probability that both will be butterscotch?**
(Skill 11.3)(Rigorous)

A)   1/2

B)   1/28

C)   1/4

D)   1/16

Answer: D

With replacement, the probability of obtaining butterscotch on the first draw is 2/8 and the probability of drawing butterscotch on the second draw is also 2/8. Multiply and reduce to lowest terms.

52) **A teacher is introducing the concept of multiplication to her third grade students. What is another way she might write 4 x 5?**
(Skill 12.1; Average rigor)

A) 4 + 5

B) 5 + 4

C) 4 + 4 + 4 + 4 + 4

D) 5 + 5 + 5 + 5 + 5

Answer: C

The multiplication concept can translate to an addition problem. 4 x 5 is the same as the number 4 added 5 times.

53) **What math principle is reinforced by matching numerals with number words?**
(Skill 12.1; Rigorous)

A)   Sequencing

B)   Greater than and less than

C)   Number representations

D)   Rote counting

Answer: C

The students are practicing recognition that a numeral (such as 5) has a corresponding number word (five) that represents the same math concept. They are not putting numbers in order (sequencing), and they are not comparing two numbers for value (greater than or less than). In this activity, students are also not counting in order just for the sake of counting (rote counting).

54)   **Line *p* has a negative slope and passes through the point (0, 0). If line *q* is perpendicular to line *p*, which of the following must be true? (Skill 12.2) (Rigorous)**

A)   Line *q* has a negative *y*-intercept.

B)   Line *q* passes through the point (0, 0)

C)   Line *q* has a positive slope.

D)   Line *q* has a positive *y*-intercept.

Answer: C

Draw a picture to help you visualize the problem.

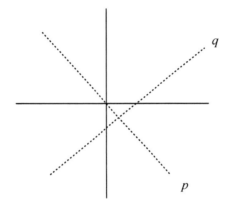

Choices (A) and (D) are not correct because line *q* could have a positive or a negative *y*-intercept. Choice (B) is incorrect because line *q* does not necessarily pass through (0, 0). Since line *q* is perpendicular to line *p*, which has a negative slope, it must have a positive slope.

55) **Cindy bought a package of cookies. She ate half of them and gave one-third of the remainder to a friend. She then had 20 fewer cookies than she did in the beginning. This information can be expressed symbolically in the following way:**
(Skill 12.3) (Rigorous)

A) $\dfrac{n}{2} - \dfrac{n}{3} = n - 20$

B) $\dfrac{n}{2} - \dfrac{n}{6} = n - 20$

C) $\dfrac{n}{2}(1 - \dfrac{1}{3}) = n - 20$

D) $\dfrac{n}{2} - \dfrac{n}{3} = n + 20$

Answer: B

If Cindy started out with n cookies, she ate $\dfrac{n}{2}$ and had $\dfrac{n}{2}$ left. Of these,

she gave her friend $\dfrac{1}{3} \times \dfrac{n}{2} = \dfrac{n}{6}$ cookies.

56) If *n* is a multiple of 3, then *n* +5 is divisible by 4.

n = 6 is a multiple of 3.
n + 5 = 6 + 5 = 11.
11 is not divisible by 4.

The above is an example of:
(Average Rigor)(Skill 13.1)

A)   Direct proof

B)   Indirect proof

C)   Counter example

D)   None of the above

Answer: C

57) Which of the following is an example of syllogism?
(Skill 13.3) (Rigorous)

A)   When it rains he carries an umbrella. It did not rain and so he did not carry an umbrella.

B)   Parrots are birds. Eagles are birds. Therefore parrots are eagles.

C)   All vegetables are nutritious. Yams are vegetables. Therefore yams are nutritious.

D)   None of the above

Answer: C

A syllogism is of the following form:
        If p, then q
        If q, then r
        Therefore if p, then r

58) **A student turns in a paper with this type of error:**
$$7 + 16 \div 8 \times 2 = 8$$
$$8 - 3 \times 3 + 4 = -5$$
(Easy)(Skill 13.4)

In order to remediate this error, a teacher should:

A)   review and drill basic number facts

B)   emphasize the importance of using parentheses in simplifying expressions

C)   emphasize the importance of working from left to right when applying the order of operations

D)   do nothing; these answers are correct

Answer: C

In the above responses, the student has shown a tendency to perform operations out of the proper order. In the first case, the student has performed the multiplication and division in a right-to-left fashion, instead of left-to-right. As a result:

$$7 + 16 \div 8 \times 2 = 7 + 16 \div 16 = 7 + 1 = 8 \qquad 8 - 3 \times 3 + 4 = 8 - 9 + 4 = 8 - 13 = -5$$

These sequences are incorrect, but when a left-to-right approach is taken, the proper answers are determined.

$$7 + 16 \div 8 \times 2 = 7 + 2 \times 2 = 7 + 4 = 11 \qquad 8 - 3 \times 3 + 4 = 8 - 9 + 4 = -1 + 4 = 3$$

Thus, emphasizing the need to work from left to right for applying the order of operations would be the correct approach to dealing with the student's pattern of errors in this case.

59) **Identify the proper sequencing of subskills when teaching graphing inequalities in two dimensions (Easy)(Skill 13.4)**

A) shading regions, graphing lines, graphing points, determining whether a line is solid or broken

B) graphing points, graphing lines, determining whether a line is solid or broken, shading regions

C) graphing points, shading regions, determining whether a line is solid or broken, graphing lines

D) graphing lines, determining whether a line is solid or broken, graphing points, shading regions

Answer: B

Graphing points is the most fundamental subskill for graphing inequalities in two dimensions. Next follows the graphing of lines, and then determining whether the line is solid or broken. The graphing of lines requires, at a minimum, the graphing of two points (such as the x- and y-intercepts). Once the line has been graphed (perhaps with a light marking), it can next be determined whether the line is solid or broken, depending on the inequality being graphed. Finally, the shading of appropriate regions on the graph may be undertaken.

60) **Which statement is incorrect? (Easy)(Skill 13.5)**

A)   Drill and practice is one good use for classroom computers.

B)   Computer programs can help to teach problem solving in the classroom.

C)   Computers are not effective unless each child in the class has his or her own workstation.

D)   Analyzing science project data on a computer during math class is an excellent use of class time.

Answer: C

It is not necessary for each student to have his own workstation for computers to be an effective classroom tool. A single computer can be sufficient when used to demonstrate some principle or idea to the class, or when computer time is divided among individual students or small groups of students. As a result, the statement in answer C is incorrect. The other answers provide legitimate statements about computers.

XAMonline, INC. 21 Orient Ave. Melrose, MA 02176
Toll Free number 800-509-4128
TO ORDER Fax 781-662-9268 OR www.XAMonline.com

GEORGIA ASSESSMENTS FOR THE CERTIFICATION OF EDUCATORS   -GACE - 2008

PO#   Store/School:

Address 1:

Address 2 (Ship to other):

City, State Zip

Credit card number_____-_____-_____-_____   expiration_____

EMAIL _____

PHONE   **FAX**

13# ISBN 2007	TITLE	Qty	Retail	Total
978-1-58197-257-3	Basic Skills 200, 201, 202			
978-1-58197-773-8	Biology 026, 027			
978-1-58197-584-0	Science 024, 025			
978-1-58197-341-9	English 020, 021			
978-1-58197-569-7	Physics 030, 031			
978-1-58197-531-4	Art Education Sample Test 109, 110			
978-1-58197-545-1	History 034, 035			
978-1-58197-774-5	Health and Physical Education 115, 116			
978-1-58197-540-6	Chemistry 028, 029			
978-1-58197-534-5	Reading 117, 118			
978-1-58197-547-5	Media Specialist 101, 102			
978-1-58197-535-2	Middle Grades Reading 012			
978-1-58197-591-8	Middle Grades Science 014			
978-1-58197-345-7	Middle Grades Mathematics 013			
978-1-58197-686-1	Middle Grades Social Science 015			
978-158-197-598-7	Middle Grades Language Arts 011			
978-1-58197-346-4	Mathematics 022, 023			
978-1-58197-549-9	Political Science 032, 033			
978-1-58197-588-8	Paraprofessional Assessment 177			
978-1-58197-589-5	Professional Pedagogy Assessment 171, 172			
978-1-58197-259-7	Early Childhood Education 001, 002			
978-1-58197-587-1	School Counseling 103, 104			
978-1-58197-541-3	Spanish 141, 142			
978-1-58197-610-6	Special Education General Curriculum 081, 082			
978-1-58197-530-7	French Sample Test 143, 144			
			SUBTOTAL	
FOR PRODUCT PRICES GO TO WWW.XAMONLINE.COM			Ship	$8.25
			TOTAL	